ELECTRICAL ENGINEERING DEVELOPMENTS

COMPRESSED SENSING

METHODS, THEORY AND APPLICATIONS

ELECTRICAL ENGINEERING DEVELOPMENTS

Additional books in this series can be found on Nova's website under the Series tab.

Additional e-books in this series can be found on Nova's website under the eBooks tab.

ELECTRICAL ENGINEERING DEVELOPMENTS

COMPRESSED SENSING

METHODS, THEORY AND APPLICATIONS

JONATHON M. SHEPPARD
EDITOR

nova
science publishers
New York

Library of Congress Cataloging-in-Publication Data

ISBN: 978-1-53613-082-9

Published by Nova Science Publishers, Inc. † New York

CONTENTS

PREFACE

Compressed Sensing: Methods, Theory and Applications presents and describes a method of image Arbitrary Sampling and Bounded Spectrum Reconstruction (ASBSR-method) that allows for drawing near the image sampling rate theoretical minimum. This compilation also discusses results of experimental verification of the ASBSR-method and its possible applicability extensions to solving various underdetermined inverse problems such as: color image demosaicing, image in-painting, image reconstruction from their sparsely sampled or decimated projections, image reconstruction from the modulus of its Fourier spectrum, and image reconstruction from its sparse samples in Fourier domain. Following this, the authors examine a novel framework to obtain HR images from CS imaging systems capturing multiple Low Resolution (LR) images of the same scene. The assumption that when an image admits a sparse representation in a transformed domain, a blurred version of it will also be sparse in the transformed domain, allows for the recovery of blurred images from CS observations. This proposed Compressed Sensing Super Resolution (CSSR) approach, combines existing CS reconstruction algorithms with an LR to HR approach based on the use of a new robust sparsity promoting prior based on super Gaussian regularisation. Additionally, several image CS recovery methods are studied. The concept of CS in image processing is introduced and a brief description of an effective CS reconstruction algorithm, called block-based CS with smoothed-projected Landweber reconstruction (BCS-SPL), is presented. Next, an adaptive CS method is presented, which provides a higher recovered image quality with respect to the BCS-SPL algorithm. Then, the concept of prediction and residual reconstruction is extended to the image CS recovery.

The problem of minimization of the number of measurements needed for digital image acquisition and reconstruction with a given accuracy is addressed in Chapter 1. Basics of the sampling theory are outlined to show that the lower bound of signal sampling rate sufficient for signal reconstruction with a given accuracy is equal to the spectrum sparsity of the signal sparse approximation that has this accuracy. It is revealed that the compressed sensing approach, which was advanced as a solution to the sampling rate minimization problem, is far from reaching the sampling rate theoretical minimum. Potentials and limitations of compressed sensing are demystified using a simple and intutive model. A method of image Arbitrary Sampling and Bounded Spectrum Reconstruction (ASBSR-method) is described that allows to draw near the image sampling rate theoretical minimum. Presented and discussed are also results of experimental verification of the ASBSR-method and its possible applicability extensions to solving various underdetermined inverse problems such as color image demosaicing, image in-painting, image reconstruction from their sparsely sampled or decimated projections, image reconstruction from the modulus of its Fourier spectrum, and image reconstruction from its sparse samples in Fourier domain.

As explained in Chapter 2, Compressed Sensing (CS) is a new technology that simultaneously acquires and compresses images reducing acquisition time and memory requirements to process or transmit them. It establishes that a sparsely representable image/signal can be recovered from a highly incomplete set of measurements or projections of the image. Image Super Resolution (SR) is an important post-processing technique where multiple input images are super resolved to obtain one or more images of higher resolution and better quality. SR necessitates a good image registration procedure in order to obtain a High Resolution (HR) image of enhanced quality. Such quality should overcome image degradation due to system hardware, optical and spatial limitations.

In Chapter 2, the authors study a novel framework to obtain HR images from CS imaging systems capturing multiple Low Resolution (LR) images of the same scene. The assumption that when an image admits a sparse representation in a transformed domain, a blurred version of it will also be sparse in the transformed domain, allows them to recover blurred images from CS observations.

Similarly, a warped, blurred, and down-sampled LR image is expected to be also sparse in a transformed domain and hence can be reconstructed from the corresponding CS observation. The proposed Compressed Sensing Super Resolution (CSSR) approach, combines existing CS reconstruction algorithms with an LR to HR approach based on the use of a new robust sparsity promoting prior based on super Gaussian regularisation.

The CSSR problem has multiple unknowns, mainly the HR image, the transformed coefficients corresponding to each LR image, and the motion vectors. The unknowns can be estimated sequentially or simultaneously. The method studied in Chapter 2 is a joint estimation framework where LR reconstructions and HR estimation are carried out simultaneously by alternating between them including the automatic estimation of registration parameters. The advantages of the joint estimation over the sequential approach and the advantage of utilising multiple input images over a single one are also discussed.

The CSSR approach converts the constrained joint optimisation problem into a sequence of unconstrained sub-problems using Alternate Direction Method of Multipliers. The HR image estimation sub-problem is solved using majorisation-minimisation. The registration parameters are estimated using two different approaches, which are compared in Chapter 2. The CS reconstruction sub-problem becomes an l_1 minimisation subject to a quadratic constraint.

The proposed framework has been tested on grayscale simulated and synthetically compressed real natural images, which were compressed using a measurement matrix that can be synthesised practically. This is compared with state-of-the-art SR algorithms which do not use CS in their work, and with state-of-the-art image deconvolution algorithms from CS images, which do not use down-sampling in their work. The proposed CSSR performs favourably compared to them. Moreover, the proposed CSSR has been tested when both compression and down-sampling are present.

Two important CSSR applications are investigated: Passive Millimeter Wave (PMMW) images and colour images. The performed CSSR experiments on synthetically compressed real millimeter wave and colour images demonstrate the capability of the CSSR framework to provide very good quality super resolved images from multiple low resolution compressed acquisitions. PMMW imaging is used for threat detection and the CSSR is expected to improve threat detection rates. For CS colour images consisting of Red, Green, and Blue (RGB) channels which were sensed separately, the registration parameters are jointly estimated from the three channels simultaneously using the CSSR framework.

Compressed sensing (CS) is a recently developed mathematical paradigm wherein the signals are sampled at the sub-Nyquist rates via a linear projection onto a random basis. Under certain conditions, the signals can be exactly recovered from these incomplete samples. In Chapter 3, several image CS recovery methods are studied. In the first part of chapter 3, the concept of CS in the image processing is introduced and a brief description of a well-known and yet effective CS reconstruction algorithm, called block-based CS with smoothed-projected Landweber reconstruction (BCS-SPL), is presented. In the next part, inspired by the fact that the visually salient regions of the image are more conspicuous to the human visual system, an adaptive CS method is presented, which provides a higher recovered image quality with respect to the BCS-SPL algorithm. Then, the authors extend the concept of prediction and residual reconstruction to the image CS recovery. A prediction is produced to create a measurement-domain residual of the image to be recovered; such a residual is typically more compressible than the original one, making it more adapted to the CS reconstruction. Several methods for the prediction generation are discussed in Chapter 3. First, considering the strong correlation among the neighboring areas within an image, the authors present an image CS recovery of images by providing an intra-image prediction. Further, the local sparsity and nonlocal similarities among the image patches are implicitly exploited to generate an optimal prediction, improving the performance of the image CS recovery. Finally, in the end part of Chapter 3, objective experimental results are given for each algorithm in terms of peak-signal to noise ratio (PSNR) and the significant gains of the presented CS recovery methods are compared to the BCS-SPL algorithm.

In: Compressed Sensing
Editor: J. M. Sheppard

ISBN: 978-1-53613-082-9

Chapter 1

COMPRESSED SENSING, THE ASBSR-METHOD OF IMAGE SAMPLING AND RECONSTRUCTION AND THE PROBLEM OF DIGITAL IMAGE ACQUISITION WITH THE LOWEST POSSIBLE SAMPLING RATE

L. P. Yaroslavsky*
Department of Physical Electronics
School of Electrical Engineering
Tel Aviv University, Tel Aviv, Israel

Abstract

The problem of minimization of the number of measurements needed for digital image acquisition and reconstruction with a given accuracy is addressed. Basics of the sampling theory are outlined to show that the lower bound of signal sampling rate sufficient for signal reconstruction with a given accuracy is equal to the spectrum sparsity of the signal sparse approximation that has this accuracy. It is revealed that the compressed sensing approach, which was advanced as a solution to the sampling rate minimization problem, is far from reaching the sampling rate theoretical minimum. Potentials and limitations of compressed sensing are demystified using a simple and intutive model. A method of image Arbitrary Sampling and Bounded Spectrum Reconstruction (ASBSR-method) is

* E-mail address: yaro@eng.tau.ac.il.

described that allows to draw near the image sampling rate theoretical minimum. Presented and discussed are also results of experimental verification of the ASBSR-method and its possible applicability extensions to solving various underdetermined inverse problems such as color image demosaicing, image in-painting, image reconstruction from their sparsely sampled or decimated projections, image reconstruction from the modulus of its Fourier spectrum, and image reconstruction from its sparse samples in Fourier domain.

Keywords: image sampling, sampling theory, sampling rate, compressed sensing

1. Introduction

Sampling is the primary operation in acquiring digital images. Contemporary digital display devices and image processing software imply by default that sampling over regular square sampling grids is used for image discretization. As it is well known, digital images acquired in this way are, as a rule, highly compressible. Hence, images are compressed for storage and transmission and then are reconstructed to the standard sampled representationfor displaying and/or processing.

The phenomenon of ubiquitous compressibility of images raises very natural questions: what are fundamental reasons of the compressibility of images sampled in the conventional way and is it possible just directly measure the minimal amount of data that won't end up being thrown away? These questions were apparently first posed by the inventors of the compressed sensing approach as a solution to this problem [1, 2, 3]. Since its introduction, the compressed sensing approach to digital image acquisition has gained a considerable popularity. It has also begotten many expectations of revolutionizing the sensor industry because it is widely believed that compressed sensing can allow high-resolution information to be recovered from low-resolution measurements made with relatively smaller detector arrays and that compressed sensing allows image sampling with sub-Nyquist sampling rates [4-8].

Potentials of compressed sensing are widely advertised in the literature. Much less is known about its limitations. Particularly important is the question: how close is the amount of measurements required by compressed sensing for image acquisition to the theoretical minimum defined by the sampling theory? The present paper discusses potentials and limitations of the compressed sensing approach from the view point of the sampling theory, shows that it requires substantially more measurements than the minimum defined by the sampling theory, discusses other limitations of compressed sensing, and describes an alternative image sampling and reconstruction method that allows to draw near the theoretical minimum of the image sampling rate.

The chapter is arranged as following. In Sect. 2 basics of the modern sampling theory are outlined. In Sect.3 sampling redundacy of compressed sensing is discussed and a simple physical interpretation of compressed sensing limited capability of reconstructing signals sampled with aliasing is offered. In Sect. 4 an Arbitrary Sampling and Bounded Spectrum Reconstruction (ASBSR-) method is described that enables sampling images with rates close to the theoretical minimum. In Sect. 5 results of experimental verification of the method are provided. In Sect. 6 some practical issues of application of the ASBSR-method are discussed. In Sect. 7 the possibilities of applying the ASBSR-method to solving various other underdetermined inverse problems are discussed. The concluding Sect.8 summarizes the results.

2. What Is the Minimal Number of Samples Required for Image Reconstruction with a Given Accuracy?

Signal sampling is based on the idea of signal band-limited approximation. For 1D signals, sampling is quite simple. Here is a signal sampling and reconstruction protocol:

- for a signal $s(x)$ to be sampled in its coordinate x, define an admissible Mean Square Error (MSE) σ^2 of its approximation;
- assuming that signal Fourier spectrum $\Im(f)=\int_{-\infty}^{\infty} s(x)\exp(i2\pi fx)dx$ is concentrated around the signal DC-component at $(f=0)$, determine a frequency interval $[-B\le f\le B]$ that contains $(E-\sigma^2)/E$ -th fraction of signal energy $E=\int_{-\infty}^{\infty}|\Im(f)|^2 df$;
- pass the signal through the ideal sampling low-pass filter *SLPF* with frequency bandwidth $[-B,B]$ to obtain its band-limited approximation $\tilde{s}(x)$:

$$\tilde{s}(x) = \int_{-\infty}^{\infty} s(\xi) SLPF(x - \xi) d\xi = \int_{-B}^{B} \Im(f) \exp(-i2\pi xf) df \qquad (1)$$

and sample this signal with sampling interval $\Delta x = 1/2B$ to obtain signal samples:

$$\tilde{s}_k = \tilde{s}(k\Delta x) = \int_{-\infty}^{\infty} s(\xi) SLPF(k\Delta x - \xi) d\xi \quad k = ...,-2,-1,0,1,2,.....; \qquad (2)$$

- For signal reconstruction, pass a comb signal $\left[\sum_{k=-\infty}^{\infty} \tilde{s}_k \delta(\xi - k\Delta x)\right]$, where $\delta(.)$ is the Dirac delta-function, through the ideal reconstruction low-pass filter *RLPF* with frequency bandwidth $[-B, B]$ to reconstruct the band-limited approximation $\tilde{s}(x)$ of the signal $s(x)$:

$$\tilde{s}(x) = \sum_{k=-\infty}^{\infty} \tilde{s}_k RLPF(x - k\Delta x). \qquad (3)$$

Point spreads functions *SLPF* and *RLPF* of sampling and reconstruction ideal low pass filters are linked by the relationship $RLPF = \Delta x SLPF$.

If a signal has a carrier frequency, say, f_0, i.e., if its Fourier spectrum is concentrated around frequency $f = \pm f_0 \neq 0$, the signal should first be demodulated by a signal $\sin(2\pi f_0 x)$ to make its spectrum concentrated around frequency $f = 0$. Then the demodulated signal can be sampled as it is described above. For reconstruction of the signal from its samples, one should first reconstruct the demodulated signal and then modulate the reconstruction result with the same modulating signal $\sin(2\pi f_0 x)$. Note that the sampling rate remains the same $2B$ no matter what is the carrier frequency f_0.

Signal sampling rate $2B$ is frequently called the "Nyquist frequency" referring to Harry Nyquist, who in 1920-th made a guess that the sufficient for signal reconstruction number of signal samples per unit time is twice the signal bandwidth. Later V. A. Kotelnikov ([9]) and C. Shannon ([10]) formulated and proved this fact as a mathematical theorem, the Kotelnikov-Shannon sampling theorem.

The Kotelnikov-Shannon sampling theorem refers to band-limited signals. In reality, signals are never band-limited. For non band-limited signals, the sampling theory states that the signal sampling rate $2B$ is the minimal sampling rate, which secures signal reconstruction with MSE equal to the signal energy outside the bandwidth $[-B, B]$. If signals are not pre-filtered before sampling by the ideal low pass filter and/or the signal reconstruction filter is not the ideal low-pass filter, which is the case in reality, aliasing artifacts appear. This why in reality signal reconstruction MSE is larger than the above minimal one.

For sampling images as 2D signals, a protocol similar to that for 1D signals is customarily used:

- for a given image $s(x, y)$ to be sampled in coordinates (x, y), choose sampling intervals $(\Delta x, \Delta y)$ (or sampling rate, in “dots per inch”) over a regular rectangular sampling grid;
- pass the image through the 2D ideal sampling low pass filter with frequency bandwidth $(-1/2\Delta x, 1/2\Delta x; -1/2\Delta y, 1/2\Delta y)$ and sample the obtained image band limited approximation $\widetilde{s}(x, y)$ with the chosen sampling intervals to obtain image samples

$$\widetilde{s}_{k,l} = \int_{-\infty}^{\infty}\int_{-\infty}^{\infty} s(x, y) SLPF(k\Delta x - x, l\Delta y - y)\,dx dy =$$

$$\int_{-1/2\Delta x}^{1/2\Delta x}\int_{-1/2\Delta y}^{1/2\Delta y} \Im(f_x, f_y)\exp[-i2\pi(k\Delta x f_x + l\Delta y f_y)]\,df_x df_y$$

$$k = ...,-2,-1,0,1,2,..... , l = ...,-2,-1,0,1,2,..... ,,, \tag{4}$$

where $\Im_s(f_x, f_y) = \int_{-\infty}^{\infty}\int_{-\infty}^{\infty} s(x, y)\exp[i2\pi(f_x x + f_y y)]\,dx dy$ is image Fourier spectrum and (f_x, f_y) are image spatial frequencies.

For image reconstruction from its samples $\{\widetilde{s}_{k,l}\}$, pass a 2D comb signal $\left[\sum_{k=-\infty}^{\infty} \widetilde{s}_{k,l}\,\delta(x - k\Delta x)\delta(y - l\Delta y)\right]$ through the ideal reconstruction low-pass filter

RLPF with frequency bandwidth $(-1/2\Delta x, 1/2\Delta x; -1/2\Delta y, 1/2\Delta y)$ to produce a band-limited approximation $\tilde{s}(x,y)$ of the image $s(x,y)$:

$$\tilde{s}(x,y) = \sum_{k=-\infty}^{\infty} \sum_{l=-\infty}^{\infty} \tilde{s}_{k,l} RLPF(x - k\Delta x, y - l\Delta y) \tag{5}$$

with approximation MSE defined by the image energy outside the band-limiting rectangle $(-1/2\Delta x, 1/2\Delta x; -1/2\Delta y, 1/2\Delta y)$ called the sampling baseband:

$$\sigma^2 = \int_{-\infty}^{\infty}\int_{-\infty}^{\infty} \left|\Im_s(f_x, f_y)\right|^2 df_x df_y - \int_{-1/2\Delta x}^{1/2\Delta x}\int_{-1/2\Delta y}^{1/2\Delta y} \left|\Im_s(f_x, f_y)\right|^2 df_x df_y =$$
$$E - \int_{-1/2\Delta x}^{1/2\Delta x}\int_{-1/2\Delta y}^{1/2\Delta y} \left|\Im_s(f_x, f_y)\right|^2 df_x df_y, \tag{6}$$

This, in fact, is how commonly used image sampling and display devices work. The role of sampling and, correspondingly, image reconstruction low-pass filters is, as a rule, played by optics of image sensors in combination with image sensor and image display device apertures. Sampling intervals $(\Delta x, \Delta y)$ are chosen usually on the base of knowledge of resolving power of imaging optics (in "lines per mm") or of visual assessment of images, assuming that one needs at least, say, 2 samples for reproducing the sharpest edges in the image.

As it was already mentioned, all image display devices and image processing software imply by default that images are sampled over square sampling grids with sampling intervals $\Delta x = \Delta y$. With such a sampling, the number N of image samples is equal to the image area S_{xy} times the area of the sampling baseband $S_\Omega = 1/\Delta x^2$: $N = S_{xy}/\Delta x^2 = S_{xy} S_\Omega$.

If all the image Fourier spectrum largest components that contain $(E - \sigma^2)/E$-th fraction of signal energy $\boldsymbol{E}$ perfectly fill a square shape of the sampling baseband $(-1/2\Delta x, 1/2\Delta x; -1/2\Delta x, 1/2\Delta x)$, $N = S_{xy}/\Delta x^2$ will be the minimal number of samples sufficient for image reconstruction with MSE σ^2. However, in reality image Fourier spectra almost never fill square shapes. This fact is illustrated in Figure 1, where 10 test images of different kind and estimates of their Fourier spectra are shown.

Fourier spectra of the images were estimated using Discrete Fourier Transform and applying to images, before spectral analysis, a circular apodization

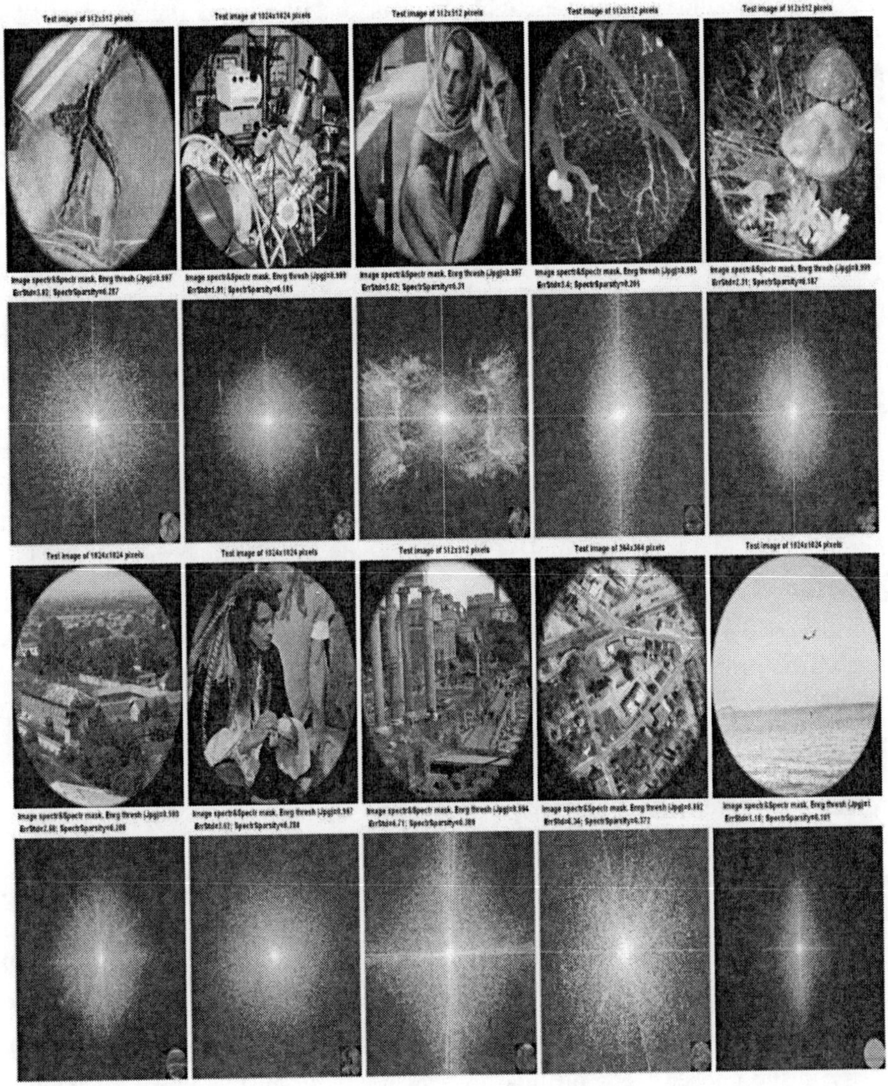

Figure 1. A set of sampled test images (first and third rows from top) and their corresponding Fourier spectra centered at spectra DC components (second and fourth rows):"Aerial Photo 512", "AFM512, "Barbara512", "Blood Vessels 512", "Mushrooms 512", "Nish 1024", "Pirat 1024, "Rome 512", "West Concord 364" and "Test 4 CS1024". Highlighted are the largest spectral components sufficient for image reconstruction with the same MSE as that of their JPEG encoding.

mask in order to smoothly bring a sampled image down to zero at the edges of the sampled region and in this way to avoid as much as possible spectrum estimation errors due to boundary effects. Highlighted in the figures of image spectra are zones of spectral components that contain the largest image spectra components sufficient for image reconstruction with the same MSE as that of their JPEG encoding. We will call them "Energy compaction" (EC-) zones. One can see in these figures that EC-zones of all images occupy only a fraction of the area of the sampling baseband. This fraction, i.e., the ratio of the area of the image EC-zone to the area of the baseband, is called image spectrum sparsity.

As it was already mentioned, the minimal sampling rate of 1D signals sufficient for signal reconstruction with a given MSE σ^2 is equal to the size $2B$ of signal spectral interval that contains $\left(E-\sigma^2\right)/E$ fraction of signal energy, i.e., to the spectrum sparsity. The same is true for images as 2D signals. In order to prove that, consider the following discrete model.

Let N be the number of image samples chosen for image sampled representation over a standard regular square sampling grid. Assume that available are $K < N$ image samples and it is required to reconstruct the entire image of N samples with minimal MSE. Choose an image transform Φ_N and use available K image samples to compute K transform coefficients. Then a "Bounded Spectrum" (BS-) approximation to the entire image can be reconstructed by the inverse transform of the set of found K transform coefficients supplemented with the rest of $N-K$ coefficients set to zero. Reconstruction MSE will be equal to the energy of $N-K$ coefficients set to zero. For a given transform, this error can be minimized if K the largest transform coefficients are used for the reconstruction. In order to further minimize image reconstruction error, one should choose a transform with a better capability of energy compaction into the smallest number of transform coefficients. The said is the meaning of the discrete sampling theorem ([14], [15]).

In terms of this discrete model, K transform coefficients that contain $\left(E-\sigma^2\right)/E$ fraction of image energy and reconstruct the image with MSE σ^2 form spectrum EC-zone of the image. The ratio $S_\sigma = K/N$ is the image spectrum sparsity in the chosen basis on this level of MSE. From the above reasoning it follows that the sampling rate (the number of samples per unit of image area) sufficient for image reconstruction with a given MSE is equal to the image spectrum sparsity on this level of MSE.

In the limit, when $N \to \infty$, the discrete model converts to a continuous one. In particular, if Discrete Fourier transform (DFT) is chosen as the image transform, it converts to the integral Fourier transform, and the discrete sampling theorem converts to the classic sampling theorem, which states that the minimal density of image samples sufficient for reconstructing images from their samples with MSE σ^2 is equal to the area B_Ω of the EC-zone Ω in the Fourier domain that contains $\left(E-\sigma^2\right)/E$ -th fraction of the image signal energy *E*. If image is sampled over a square sampling grid with sampling intervals Δx, the square baseband $\left(-1/2\Delta x, 1/2\Delta x; -1/2\Delta x, 1/2\Delta x\right)$ must encompass the image spectrum EC-zone. Therefore, the number of image samples on a square sampling grid $N = S_{xy}/\Delta x^2$ will always inevitably exceed the minimal required number of samples $N_{\min} = S_{xy}B_\Omega$. This is the reason for the ubiquitous compressibility of digital images. Relative, with respect to the square baseband, areas of EC-zones of test images shown in Figure 1, i.e., sparsities of their DFT spectra, range between 0.1 and 0.39, which means that the images are from 2.5 to 10 times oversampled.

3. Compressed Sensing: Why and When Is It Possible to Precisely Reconstruct Signals Sampled with Aliasing?

Compressed sensing approach is based on the idea of signal sparse approximation, i.e., approximation of signals by signals whose spectrum in a certain chosen "sparsifying" transform contains less non-zero transform coefficients than the total number of the coefficients. It was proven in the theory of compressed sensing that if an image of *N* samples is known to have, in the domain of a certain transform, only *K* non-zero transform coefficients out of *N*, the image can be precisely reconstructed from *M* > *K* measurements by means of minimization of L0 or, which is more practical, L1 norm in the image transform domain, provided the following inequality holds ([11]).

$$M/K > -2\log[(M/K)(K/N)] \tag{7}$$

As it was shown in the previous section, signal sparsity $Ss = K/N$ is the theoretical minimum of the sampling rate required for signal reconstruction. Therefore, the ratio $R = M/K$ of the number of required measurements *M* to the number *K* of signal non-zero transform coefficients represents sampling redundancy with respect to the theoretical minimum. Inequality (7) can be rewritten as a relationship between signal sparsity $Ss = K/N$ and sampling redundancy $R = M/K$ as

$$R > -2\log(R \times Ss) \tag{8}$$

Numerical evaluation of the relationship (8) between sampling redundancy $R = M/K$ and signal sparsity $Ss = K/N$ is presented in Figure 2 by dash-dot line. Solid line in this figure is plotted on the base of experimental data reported in the literature ([12]).

One can see in Figure 2 that, in the range of sparsities of test images shown in Figure 1 from 0.1 to 0.39, sampling redundancy of compressed sensing should theoretically be larger than 2 – 3 and, according to the experimental curve, it should be larger than 2.5 – 5. This means that sampling redundancy of compressed sensing for this set of test images is not much lower than that of their regular sampling (2.5 - 10) and that compressed sensing is quite far from reaching the theoretical minimum of signal sampling rates.

The mechanism of image acquisition and reconstruction by means of the compressed sensing approach and the nature of its sampling efficiency limitation given by Eq. 8 are in the current literature, as a rule, hidden in thickets of mathematical theorems. They can be demystified on the following simple example.

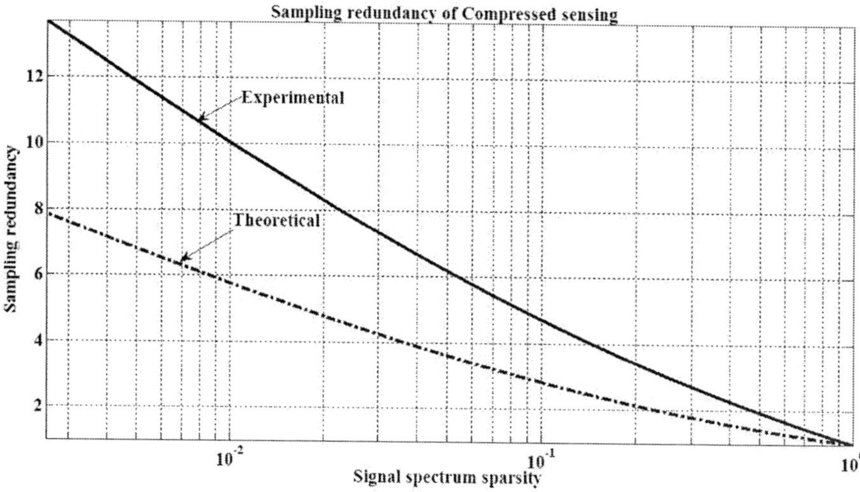

Figure 2. Theoretical and experimental relationships between signal sparsity and sampling redundancy required by the compressed sensing approach. For sampling redundancies above the curves signal reconstruction is possible, otherwise it is not.

Let a signal of N samples composed of a certain known number $K<N$ of sinusoidal components be subsampled arbitrarily in $M<N$ points and it is required

to precisely reconstruct the signal from these M samples, i.e., to determine amplitudes and frequencies of the signal sinusoidal components. Figure 3 illustrates how and when this can be done. Plots in this figure are numbered from top to bottom.

A test signal shown in Figure 3, the 1st plot, is composed of three sinusoidal components (K=3, N=256) seen as three Kronecker deltas in the signal Discrete Cosine Transform (DCT) spectrum (the 2nd plot in Figure 3). When this signal is subsampled as it is shown in the 3rd plot in Figure 3, aliasing spectral components appear in the spectrum of the subsampled image. They can be seen in the 4th plot in Figure 3. In our example the signal is sub-sampled in random positions with sampling rate 0.15 (M=38).

Intensity of the aliasing components grows with lowering the subsampling rate. Obviously, however, that, while the subsampling rate is not too low, the true spectral components of the signal exceed the highest picks of the aliasing spectral components and one can detect those true components by means of finding positions of the given number K (in this particular case K=3) the largest spectral components (the 4th plot in Figure 3). Ones these positions are determined, the following iterative Gerchberg-Papoulis type signal reconstruction algorithm can be run:

- Compute DCT of the current estimate of the reconstructed signal.
- Detect the given number K the largest spectral components.
- Set to zero all spectral components except the detected ones.
- Compute inverse DCT of the modified in this way signal spectrum to get a next estimate of the reconstructed signal.
- Replace samples of the obtained estimate of the reconstructed signal in the positions of available signal samples by the available ones.
- Repeat the loop.

The plot of reconstruction RMSE vs the number of reconstruction iterations (the 5th plot in figure 3) and plot of the reconstructed signal DCT spectrum (the 6th plot) illustrate this process and show that virtually precise reconstruction of the signal is achieved after a couple of tens of iterations: after 25 iterations reconstruction RMSE is 3.8×10^{-5}. Note that in this particular example signal sparsity is $Ss = 3/256 \approx 1.2 \times 10^{-2}$ and sampling redundancy is $R = M/K = 38/3 \cong 12.7$.

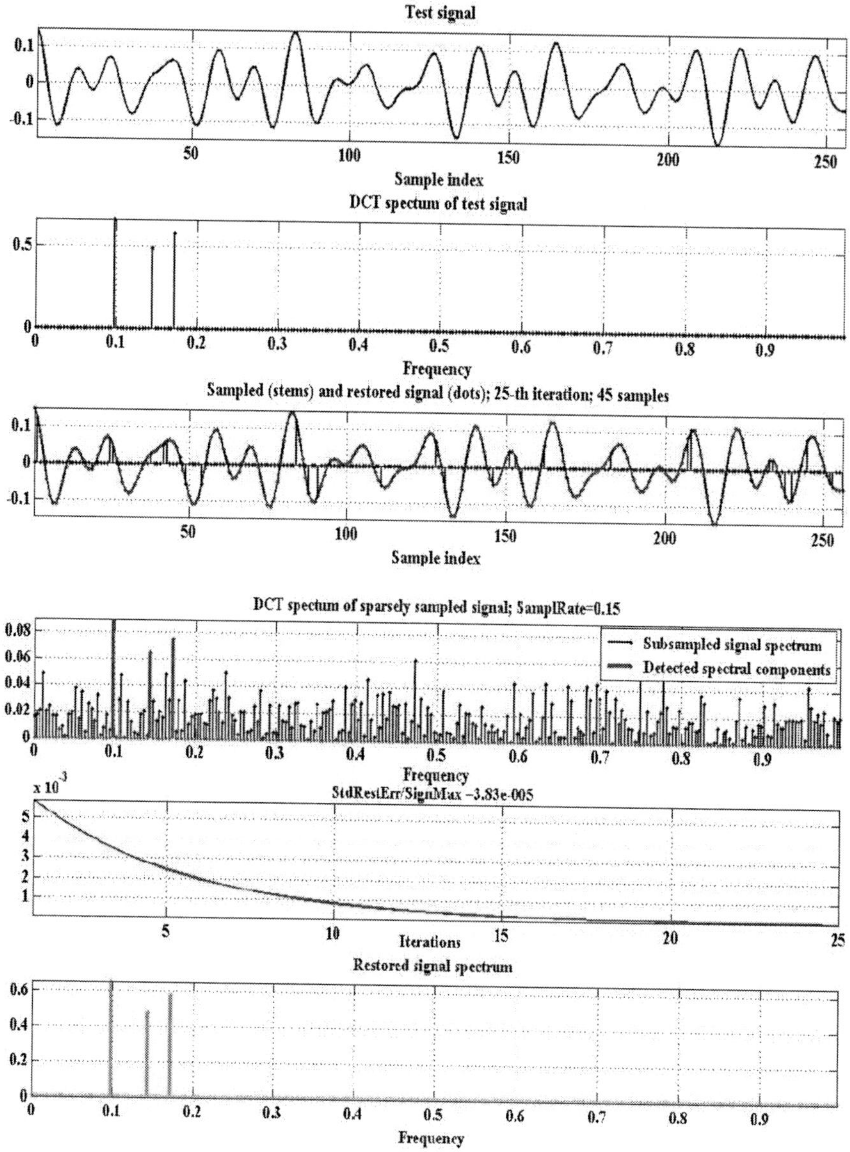

Figure 3. From top to bottom: test signal composed of 3 sinusoidal components; its DCT spectrum; this signal randomly subsampled and reconstructed; DCT spectrum of the subsampled signal; plot of reconstruction root mean square error (RMSE) vs the number of reconstruction iterations; DCT spectrum of the reconstructed signal. Frequency in plots of spectra is given in fraction of the sampling baseband.

If signal subsampling rate is too low, and aliasing is severe, reliable detection of the signal spectral components in the spectrum of the sub-sampled signal and, therefore, signal reconstruction become impossible. It is clear also that the presence of noise in the sampled data hampers reliable detection of the signal spectral components and requires additional sampling redundancy for reliable signal reconstruction.

When subsampling is carried out in random positions, one can evaluate performance of the described method of signal reconstruction in terms of the probability of error in detecting signal spectral components. This probability depends on the sub-sampling rate and on the signal sparsity K/N.

Figure 4, left plot, presents results of experimental evaluation of the probability of frequency identification error as a function of sub-sampling rate for a sinusoidal signal (K=1) of 5 different frequencies (0.9, 0.7, 0.5, 0.3, and 0.1 of the signal baseband), 5 different signal lengths N(128, 256, 512, 1024, 2048), and, correspondingly, of 5 different signal sparsities K/N. These results for each value of signal sparsity are averages over the results obtained for different frequencies. Right plots in Figure 4 show sampling redundancies required for signal reconstruction by the above described algorithm with probability of signal frequency identification error less than 10^{-4}, 10^{-3}, and 10^{-2}. The results were obtained by a Monte-Carlo simulation of the algorithm with 5×10^4 realizations of random sampling for each individual experiment with a given sampling rate, signal frequency, and signal length.

These results illustrate the principles of the compressed sensing approach to reconstruction of signals sampled with aliasing. They clearly demonstrate why a considerable sampling redundancy is required in order to secure reliable reconstruction of such signals using a priori knowledge of the signal sparsity.

The sampling redundancy required by compressed sensing is not its only drawback. The applicability of compressed sensing is also impeded by its vulnerability to noise in sensed data and by the impossibility to predict and secure the resolving power of reconstructed images. Resolving power of images is determined by the size and shape of EC-zones of their spectra. EC-zones of spectra of images reconstructed by methods of compressed sensing are formed in the process of image reconstruction rather than are specified in advance from the requirements to image resolving power.

To summarize, compressed sensing methods are to a certain degree capable of reconstructing sparse approximations of images sampled with aliasing. No knowledge regarding EC-zones of image spectra is required for this. One has only

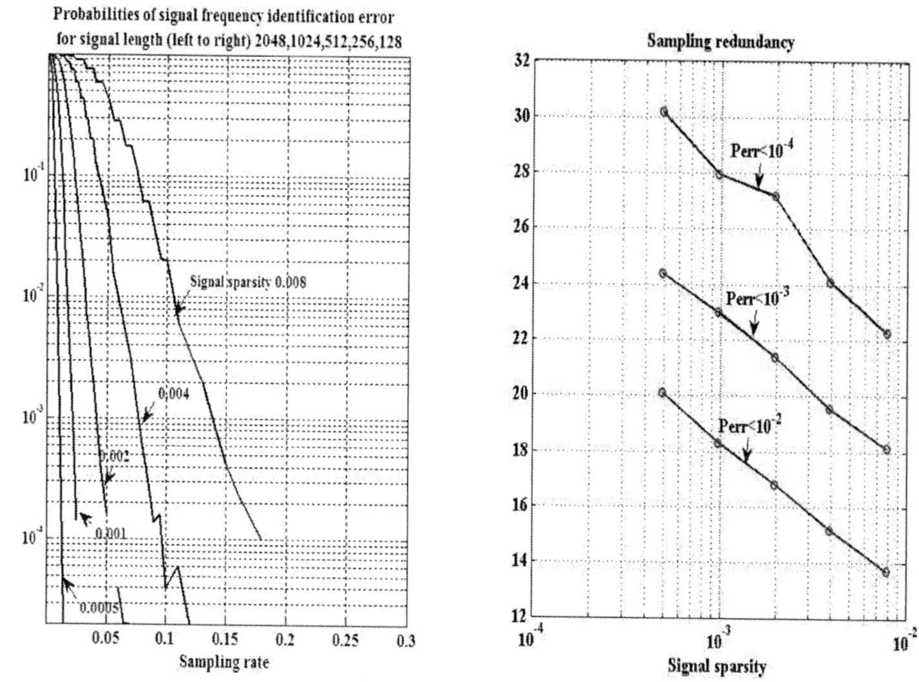

Figure 4. Plots of the probabilities of error in identification of signal frequency (left) and estimates of sampling redundancy as a function of signal sparsity K/N(right), for K=1 and N= 128, 256, 512, 1024, 2048.

to choose an image sparsifying transform. Not using any knowledge regarding image spectra EC-zones has its price. Because of this, compressed sensing requires a significant redundancy in the number of sufficient for image reconstruction measurements compared to the theoretical minimum.

In many practical tasks of digital image acquisition, the assumption of the complete uncertainty regarding image spectra EC-zones has no justification. In fact, if one is ready, as it is supposed by the compressed sensing approach, to accept a sparse approximation to an image and has chosen an image sparsifying transform, one tacitly implies certain knowledge of the energy compaction capability of the chosen transform. In what follows we show that making use of this in any way available a priori knowledge allows implementing image sampling with rates close to the theoretical minimum ([13]).

4. ASBSR-Method of Image Sampling and Reconstruction

As it was outlined in Sect. 2, sampling theory suggests the following principle of image sampling and reconstruction:

- Choose the required number N of image samples.
- Choose an image sparsifying transform.
- Specify a desired EC-zone of the image spectrum, i.e., a set of $M \leq N$ transform coefficients to be used for image reconstruction.
- Measure M image samples.
- Use M image samples for determining M transform coefficients of the chosen EC-zone.
- Set the rest N - M transform coefficients to zero and use the obtained spectrum for reconstruction of required N image samples by means of its inverse transform.

Consider possible ways for implementing this principle.

Choosing a Transform

The choice of the transform is governed by the transform energy compaction capability. An additional requirement is the availability of a fast transform algorithm. From this point of view, Discrete Cosine Transform (DCT), Discrete Fourier Transform (DFT) and wavelet transforms are among primary candidates.

Specification of Image EC-Zones

Specification of the subset of image transform coefficients to be used for image reconstruction, i.e., of the image spectrum EC-zone, can be made on the basis of the known capability of image transforms, such as DCT, to compact most of the image signal energy into few transform coefficients. In what follows we will assume using DCT as the image sparsifying transform. DCT is known to compact the image largest transform coefficients into more or less compacted groups in the area of coefficient's lower indices around the DC component.

Considerable practical experience, including that obtained in course of developing of zonal quantization tables for JPEG image compression standard, shows that although these groups of DCT coefficients do not have sharp borders, they are quite well concentrated. This means that the groups can be, with a reasonably good accuracy in terms of preservation of the group total energy, circumscribed by one of some standard shapes that encompass area of image low spatial frequencies and can be specified by few parameters, such as area, aspect ratio, angular orientation, etc.

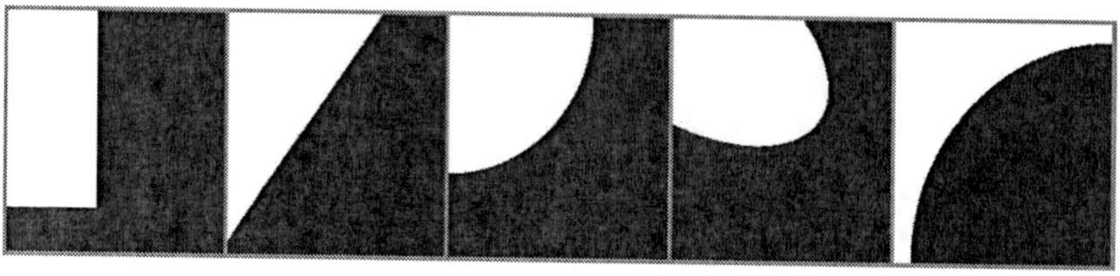

Figure 5. Examples of possible standard shapes of image EC-zones for image DCT spectra. From left to right: rectangle, triangle, pie-sector, oval, super ellipse. Spectrum DC components are in the upper left corners of the shapes.

Figure 5 presents a set of five possible standard shapes suited for DCT as the sparsifying transform: rectangle, triangle, pie-sector, ellipse, and super ellipse. In principle, each particular standard shape can be associated with a certain class of images, such as micrographs, aerial photographs, space photos, in-door and out-door scenes, etc.

The author's experimental experience shows that no fine tuning of shape parameters is required for specifying chosen shapes as approximations of image EC-zones. This property of sparse DCT spectra is illustrated in Figure 6 on an example of sparse DCT spectrum of test image "BloodVessels512".One can easily notice in the image a certain preferential prevalence of horizontal edges. This causes anisotropy of image sparse spectrum seen in boxes b) – d) in Figure 6, where shown are, marked as white dots, the largest DCT coefficients that reconstruct this image with root mean square error (RMSE) 3.85 gray levels of 255 levels, the same as the reconstruction RMSE of this image after its standard JPEG compression by the Matlab tools. These coefficients occupy 0.164 of the image baseband area and form the image spectrum EC-zone. Additionally in boxes b) – e) shown are borders of rectangular, triangular and oval shapes that all have the same area (0.275 of the baseband area) and different aspect ratios (0.35, 0.25, 0.3, and 0.45, correspondingly). When used as spectrum bounding shapes that approximate the image EC-zone, they all reconstruct the test image with practically the same reconstruction RMSEs (4.1, 3.7, 3.8, and 3.8 of image gray levels, correspondingly) as that of the JPEG compression (3.85).

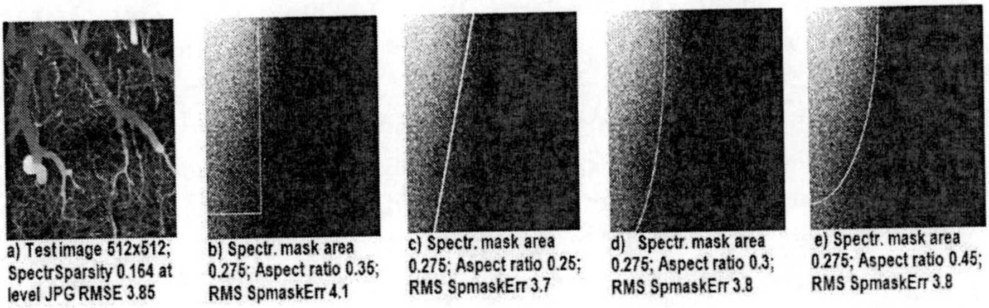

Figure 6. Test image "Blood Vessels 512" with spectrum sparsity 0.164 at the level of JPEG encoding RMSE 3.85 of image gray levels (a), positions of the image largest spectral components that reconstruct image with RMSE 3.85 (white dots), and borders (white lines) of approximate rectangular, triangular and oval shapes with different shape parameters (b) –e)).

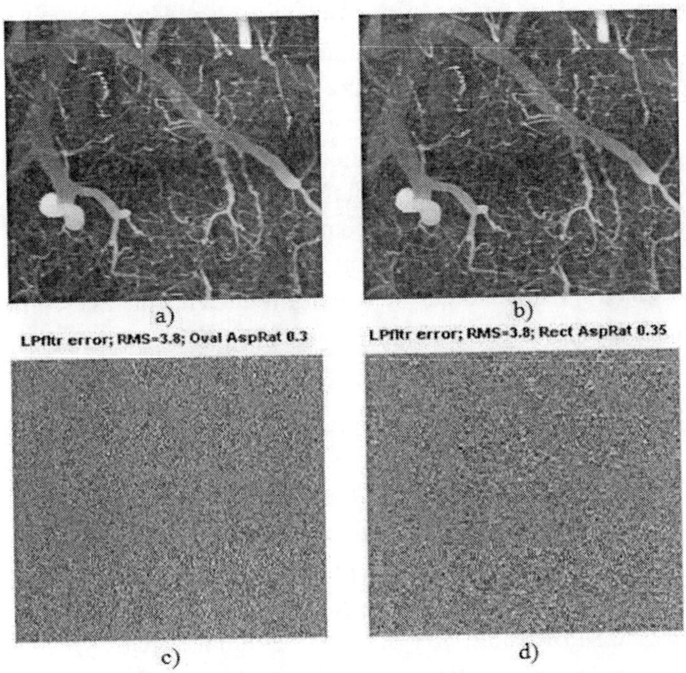

Figure 7. a), b) - Images reconstructed after bounding DCT spectrum of the test image of Figure 6, a) by oval (Figure 6, d) and by rectangular (Figure 6, b) spectral masks; c), d) - patterns of the corresponding reconstruction errors (displayed 8 times contrasted).

Images with spectra bounded by one of these shapes are visually indistinguishable one from another, though patterns of the reconstruction errors look, naturally, a bit different. Two examples of such images obtained for spectrum bounding by the rectangle, as in Figure 6, b), and by the oval, as in Figure 6 d), are shown in Figure 7, a) and b) along with corresponding patterns of the reconstruction errors (Figure 7, c) and d)). For display purposes, reconstruction errors are shown 8 times contrasted.

As one can see in Figure 6, shapes that are chosen to approximate the image spectrum EC-zone do not include all spectral coefficients of the EC-zone. From the other side, they include some of spectrum components that do not belong to the EC-zone. Those components that do not belong to the EC-zone and happen to be inside the chosen spectral shape have, by definition, lower intensity than components of EC-zone that happen to be outside the shape. Therefore, given the energy of all spectral components encompassed by the chosen spectral shape, the number of these internal "no-EC-zone components" must exceed the number of EC-zone components not encompassed by the shape. This means that the area of the shape that defines the number of samples to be taken will always exceed to a certain degree the number of EC-zone coefficients, which, theoretically, is the minimal number of samples required. For instance, redundancies of approximating EC-zone shapes of the image in Figure 6 are 0.275/0.164 = 1.67. Experimental results reported in the next section show that redundancies of standard EC-zones for natural images are, as a rule, of the same order of magnitude. This EC-zone shape's redundancy is the price one should pay for not knowing exact indices of transform coefficients that form the image spectrum EC-zone.

Specifying Positioning of Image Samples

Positioning of image samples should permit computation from them transform coefficients chosen for image reconstruction. Some image transforms, such as wavelets, impose in this respect certain limitation on positions of image samples. For DFT and DCT, positions of image samples can be arbitrary ([14]) An additional advantage of using DFT and DCT as image sparsifying transforms is that they are discrete representations of the integral Fourier transform and, as such, they ideally concord with characterization of imaging systems in terms of their Modulation Transfer Functions.

Methods of Image Reconstruction

For image reconstruction, one can consider two options:

- Direct inversion of the $M \times N$ inverse transform matrix that links M available samples and M transform coefficients specified by the chosen spectral EC-zone with the rest N - M coefficients set to zero. Once M chosen transform coefficients are found and the rest N - M transform coefficients are set to zero, the inverse transform is applied to the found spectrum to reconstruct all required N samples. Generally, matrix inversion is a very hard computational task and no fast matrix inversion algorithms are known. In our particular case, a pruned fast transform matrix should be inverted. There exist pruned versions of fast transforms for computing subsets of transform coefficients of signals with all samples except several ones equal to zero [15] Yaroslavsky L., *Theoretical Foundations of Digital Imaging*, CRC Press, Roca Baton, [16], which is inverse to what is required in the given case. The question, whether these pruned algorithms can be inverted for computing a subset of transform non-zero coefficients from a subset of signal samples, is open.

- An iterative Papoulis-Gerchberg type algorithm. The algorithm at each iteration consists of two steps:

  (i) The iterated reconstructed image is subjected to the direct transform and then spectral coefficients outside the chosen bounding EC-zone are zeroed for obtaining an iterated image spectrum.

  (ii) The iterated image spectrum is inverse transformed and then samples of the obtained image at positions, where they were actually taken at sampling, are replaced by the corresponding available ones, which produces the next iterated reconstructed image. As a zero order approximation, from which the reconstruction iterations start, an image interpolated in one or another way from the available samples can be taken. A particular interpolation algorithm used in verification experiments is detailed in the next section.

The above reasoning suggests the following protocol of image sampling and reconstruction with DCT as the image sparsifying transform:

Image sampling:

- Choose the required image spatial resolution SpR (in "dots per inch") in the same way as it is being done in the conventional image sampling.
- On the basis of evaluating the image, choose one of the standard spectral bounding shapes for bounding EC-zone of DCT spectrum and their shape parameters, such as aspect ratio for rectangle and triangle, aspect ratio and orientation angle for ellipse, etc.
- Evaluate X and Y dimensions $ShSzX$ and $ShSzY$ of the chosen shape using SpR as the shape largest diameter.
- Specify the number of pixels $N_x \geq ShSzX$ and $N_y \geq ShSzY$ "per inch" in X and Y dimensions of the reconstructed image.
- For the chosen shape, evaluate the fraction $\boldsymbol{Fr}$ of the area the shape occupies in the rectangle $N_x \times N_y$. This fraction times $SpR \times SpR$ determines spatial density $SpD=Fr \times SpR^2$ of samples to be taken (in "dots per square inch"). The number of samples M to be taken can be then found as a product of SpD and the image area $ImgSzX \times ImgSzY$: $M = SpD \times ImgSzX\ ImgSzY$.
- Choose whatever sampling grid appropriate for the available image sensor and sample the image in M positions; if no other option is available, use sensor's aperture as a pre-sampling low-pass filter.

Image reconstruction:

- Apply to the sampled image one of the above-described reconstruction options using for specification of the image spectrum EC-zone the chosen spectrum bounding shape. In this way, an image with spectrum in the chosen transform bounded by the chosen EC-zone, or a bounded spectrum (BS-) image, will be obtained, which has the prescribed spatial resolution SpR.

As one can see, the described sampling protocol does not essentially differ from the ordinary standard 2D sampling protocol. The only difference is that arbitrary sampling grids can be used and evaluation of image expected spectrum shape for approximating image spectrum EC-zone is required in the suggested method in addition to the specification of the desired image resolution, which anyway is required by the standard sampling protocol.

Not much different is image reconstruction from sampled data as well. In the suggested method, low pass filtering at image reconstruction is carried out numerically by means of bounding image spectrum in the chosen transform by the chosen spectral shape. Thanks to this, the method reaches the minimal sampling rate defined by the area of the chosen spectral shape. As it was mentioned, the latter is somehow larger than the area of the image sparse spectrum, which it approximates and which defines the absolute minimum of the image sampling rate.

In view of the said the suggested image sampling and reconstruction method can be called Arbitrary Sampling and Bounded Spectrum Reconstruction (ASBSR-) method.

5. Experimental Verification of the Method

The suggested image ASBSR-method has been experimentally verified on a considerable data base of test images including ten images presented in Figure 1. In the experiments, the above-described iterative Gerchberg-Papoulis type algorithm was used and three types of sampling grids were tested:

- "quasi-uniform" sampling grid, in which M image samples are distributed uniformly with appropriate rounding off their positions to the nearest nodes of the dense square sampling grid of N samples;
- uniform sampling grid with jitter, in which horizontal and vertical positions of each of M samples are randomly chosen, independently in each of two image coordinates, within the primary uniform sampling intervals;
- totally pseudorandom sampling grid, in which positions of samples are uniformly distributed in a pseudo-random order over nodes of the dense sampling grid of N samples.

As an image transform that compacts the image spectrum, the Discrete Cosine Transform was used. As an admissible RMSE of approximation of test images by images with sparse DCT spectra, RMSE of image compression by the standard

JPEG compression in Matlab implementation is taken. These RMSEs were used for choosing areas of EC-zone shapes for each particular image. The chosen shapes were used for bounding image DCT spectra both for image pre-filtering before sampling and in the process of image reconstruction. The former is important not only for avoiding aliasing artifacts but also for securing convergence of the iterative reconstruction algorithm to an image with spectrum bounded by the chosen EC-shape.

As a starting zero-order approximation to reconstructed images, each not available image sample was interpolated from three nearest to it available samples taken with weights inversely proportional to their distances from the interpolated sample.

Figure 8, Figure 9 and Figure 10 illustrate results of experiments with ten images of the tested set. Shown in Figure 8are: (i) test image, (ii) reconstructed image, (iii) sampled test image, (iv) the border of the chosen shape of image EC-zone (solid line) and positions of the image DCT spectrum largest coefficients (white dots), which reconstruct image with RMSE equal to that of image JPEG compression; (v) plots of RMS of all reconstruction errors (difference between the initial test image pre-filtered for bounding its spectrum before sampling and the reconstructed image) and of RMS of the smallest 90% of reconstruction errors vs the iteration number. Separate counting RMS of the smallest 90% of reconstruction errors is motivated by the observation that the iterative reconstruction converges not uniformly over the image area: most of the errors decay with iterations much more rapidly than few isolated large errors. Reconstruction RMSE are given in units of image gray levels in the range 0-255.

In Figure 9 and Figure 10 shown are, for the sake of saving space, only (i) reconstructed images (left column), (ii) positions of the largest DCT coefficients of sparse approximations to the corresponding test images and borders of their chosen EC-zones(middle column), and (iii) plots of reconstruction RMSE versus the number of iterations (right column).

For all shown images, sampling over uniform sampling grids with jitter was used, for which reconstruction errors decayed with iterations most rapidly. For the same number of iterations, reconstruction RMSE for totally random sampling grids was about 1.5-2 times and for "quasi-uniform" sampling grids 2-2.5 times larger than that for the "uniform with jitter" sampling grid. For "quasi-uniform" sampling grids, stagnation of the iteration process was observed, which can apparently be attributed to the presence of regular patterns of thickenings and rarefications of sampling positions due to rounding off their coördinates to positions of nodes of the regular uniform sampling grid.

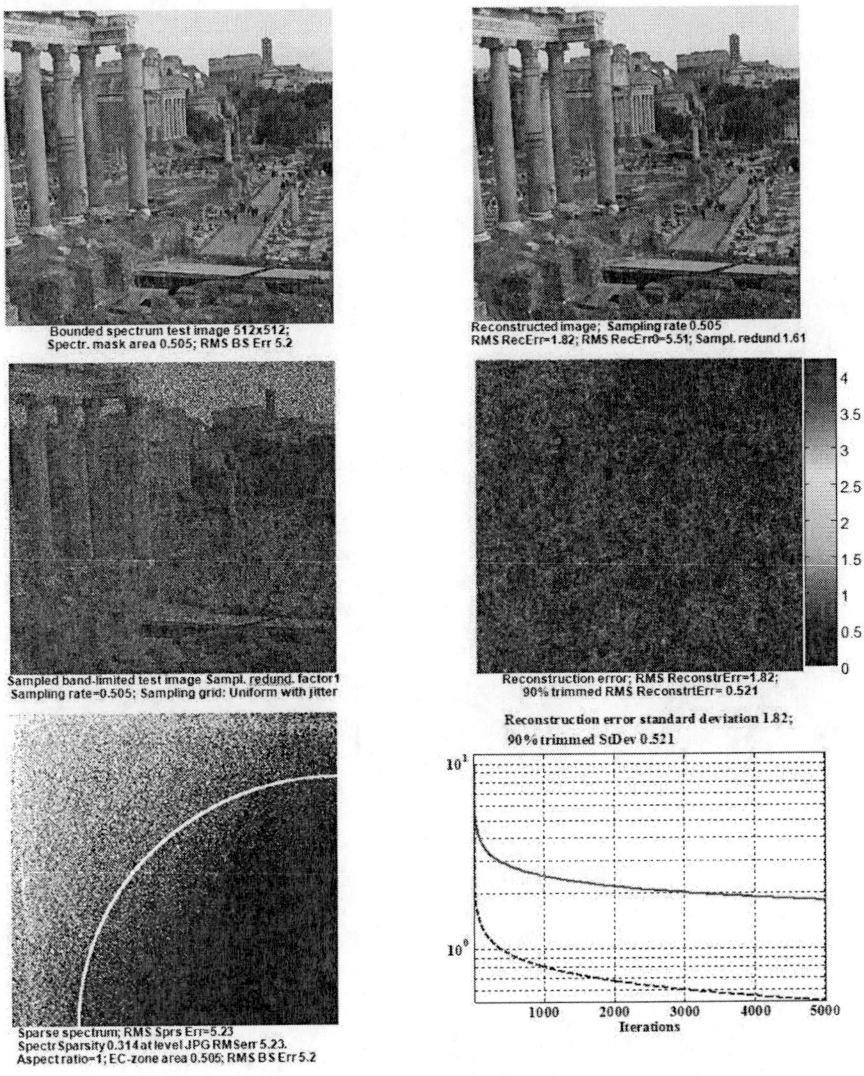

Figure 8. Results of experiments on sampling and BS-reconstruction of test image "Rome512". From top to bottom, from left to right: test image; reconstructed BS-image; sampled test image; test image sparse spectrum (white dots) and the border of the chosen EC-zone (white solid line); color coded (Matlab color map "jet", color on-line) absolute value of reconstruction error (difference between initial and reconstructed images); plots of RMS of all (solid line) and of the smallest 90% (dash line) of reconstruction errors vs the number of iterations.

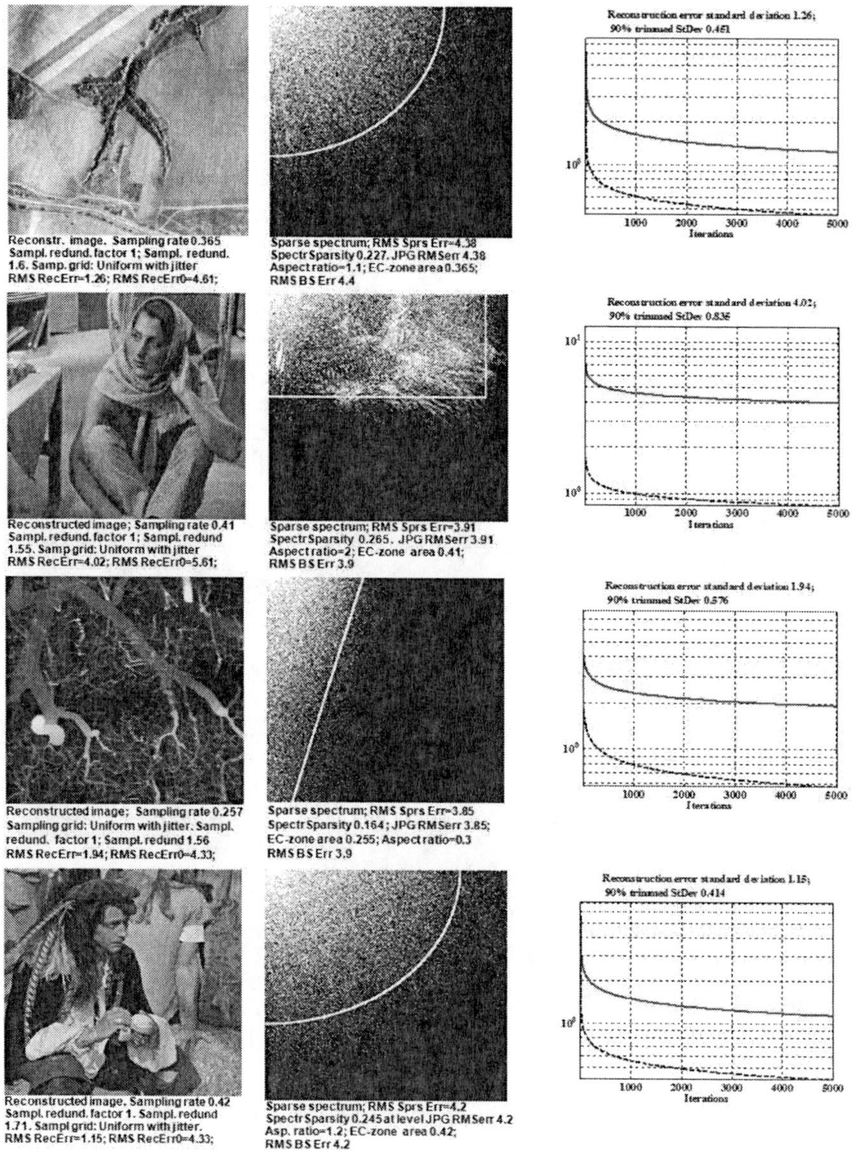

Figure 9. Results of experiments on sampling and reconstruction of test images. From top to bottom: “AerialPhoto512” “Barbara512”,“BloodVessels512”, “Pirat1024”. From left to right: reconstructed images; image sparse spectra (white dots) and borders of the corresponding chosen EC-zones (white solid line); plots of RMS of all (solid line) and of the smallest 90% (dash line) of reconstruction errors vs the number of iterations.

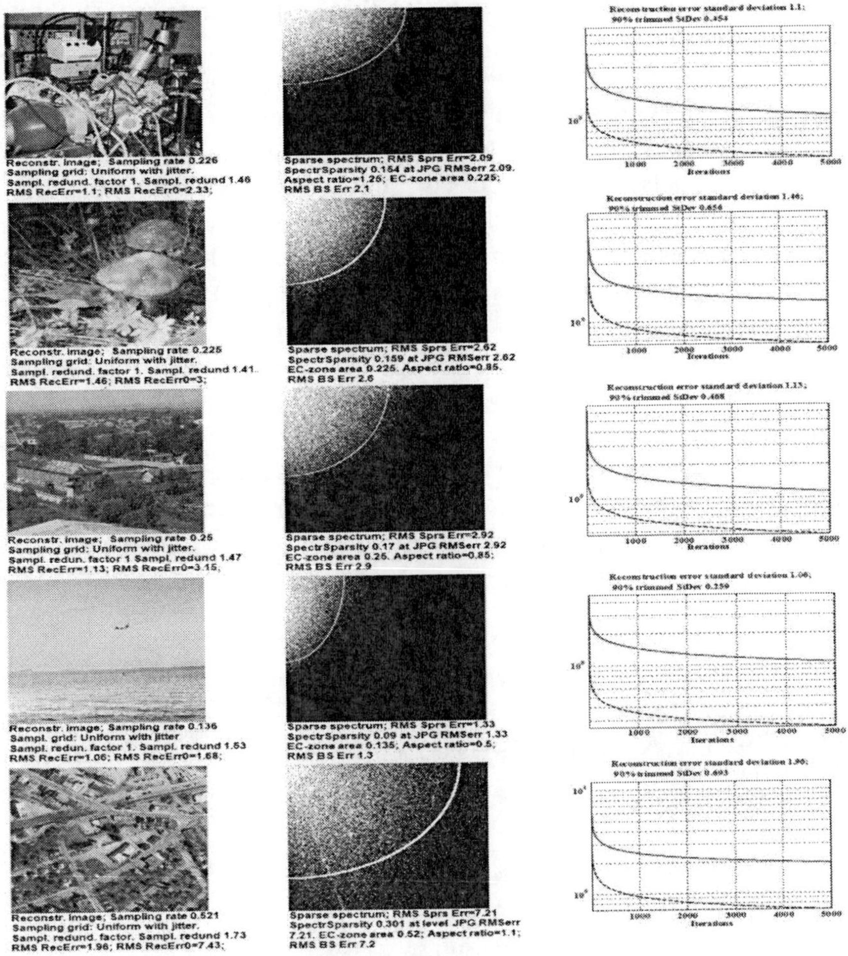

Figure 10. Results of experiments on sampling and reconstruction of test images. From top to bottom: “AFM 1024”, “Mushrooms 512”, “Nish 1024”, “Test 4 CS1024”, “West Concord 364”. From left to right: reconstructed images; image sparse spectra (white dots) and borders of the corresponding chosen EC-zones (white solid line); plots of RMS of all (solid line) and of the smallest 90% (dash line) of reconstruction errors vs the number of iterations.

Numerical results for reconstruction accuracy, spectrum sparsity, sampling rate, redundancy of the chosen EC-zones (the ratio of fractions of area they occupy in the sampling baseband to the spectrum sparsity), sampling redundancy (the ratio of the sampling rate to the relative area of the chosen EC-zone), and the overall sampling redundancy (the ratio of the sampling rate to the spectrum sparsity) presented in Table 1.

Plots of reconstruction RMSEs vs the number of iterations in Figure 8, Figure 9 and Figure 10 show that RMS of reconstruction errors decay at the first couple of hundreds iterations quite rapidly but after they reach the value of about 2-3 quantization intervals, the error decay slows down. It was found in the experiments that one can substantially accelerate the error decay, if the number of samples is taken with a certain small redundancy, i.e., 10-20% larger than the minimal number equal to the area of the chosen EC-zone (see results for test image "Barbara 512" in Table 1).

To summarize, the experiments confirm that images sampled with sampling rates equal to the minimal rate for their chosen EC-zones can be reconstructed with a sufficiently good accuracy. The redundancy in the number of required samples associated with that of standard shapes of image spectra EC-zones is of the order 1.5-1.6 and it never exceeded 2 in experiments with other images. These figures are estimates of the sampling redundancy of the suggested ASBSR-method.

Table 1. Summary of experimental results

Test image	Reconstruction RMSE (Peak signal to RMSE ratio)	Spectrum sparsity	Sampling rate	Sampling redundancy		
				Redundancy of chosen EC-zone	Sampling redundancy with respect to chosen EC-zone	Overall sampling redundancy
Aerial Photo 512	1.26 (46.2 dB)	0.227	0.365	1.6	1	1.6
AFM 1024	1.1(47.3 dB)	0.154	0.226	1.46	1	1.46
Barbara 512	4.02 (36.1 dB)	0.265	0.412	1.55	1	1.55
	0.69 (51.4 dB)	0.265	0.474	1.55	1.15	1.78
Blood Vessels 512	1.94 (42.4 dB)	0.164	0.257	1.56	1	1.56
Mushrooms 512	1.46 (44.9 dB)	0.159	0.226	1.41	1	1.41
Nish 1024	1.13 (47.1 dB)	0.17	0.25	1.47	1	1.47
Pirat 1024	1.15 (46.9 dB)	0.245	0.42	1.61	1	1.61
Rome 512	1.92 (42.5 dB)	0.314	0.5	1.59	1	1.59
Test 4 CS1024	1.06 (47.7 dB)	0.09	0.136	1.53	1	1.53
West Concord	1.96 (42.3 dB)	0.314	0.521	1.73	1	1.73

6. Some Practical Issues

In this section, four issues of practical application of the suggested image ASBSR- method are addressed: (i) how robust is the method to the presence of noise in sensor data; (ii) practical considerations regarding choosing the shape of EC-zone for bounding image spectra for image sampling and reconstruction, (iii) image antialiasing pre-filtering, and (iv) computational complexity of the method.

Choosing the Shape of EC-Zone for Bounding Image Spectra for Image Sampling and Reconstruction

As it was mentioned in Sect. 4, no fine tuning is required for specifying shape parameters. Therefore, it is suggested that several standard shapes, such as those shown in Figure 5, should be found for different classes of images, e.g., landscape, portrait, micrographs, aerial and space photographs of different kind, and alike. This can be based, for instance, on a machine learning algorithm trained on various image data bases. For sampling images in a particular application, the user should only specify an image class. Note that specifying an image class is the standard option for setting parameters of modern digital cameras.

Noise Robustness of the ASBSR-Method

From the method description in Section 4, one can see that the ASBSR-method, just as the conventional sampling and reconstruction, is linear, i.e., it satisfies the superposition principle. No parameter of sampling and reconstruction algorithms depends on signal values and, in particular, on whether noise is present in the signal or not. Therefore, sampling and reconstruction of an image that contains additive noise will result in a reconstructed image that also contains additive noise with power spectrum bounded by the shape of the spectrum EC-zone used for image reconstruction. In particular, if the sensor noise with variance σ^2 has uniform power spectrum within the sampling baseband, noise in the reconstructed image will have variance $\kappa\sigma^2$, where $\kappa < 1$ is the relative area of the reconstruction EC-zone, and its power spectrum will be uniform within the spectrum EC-zone and zero outside it.

In order to illustrate the said, an experiment on sampling and reconstruction of an image with and without additive noise was conducted. The results are presented in Figure 11, from which one can see that the presence of noise in sampled data has no influence on the work of the reconstruction algorithm.

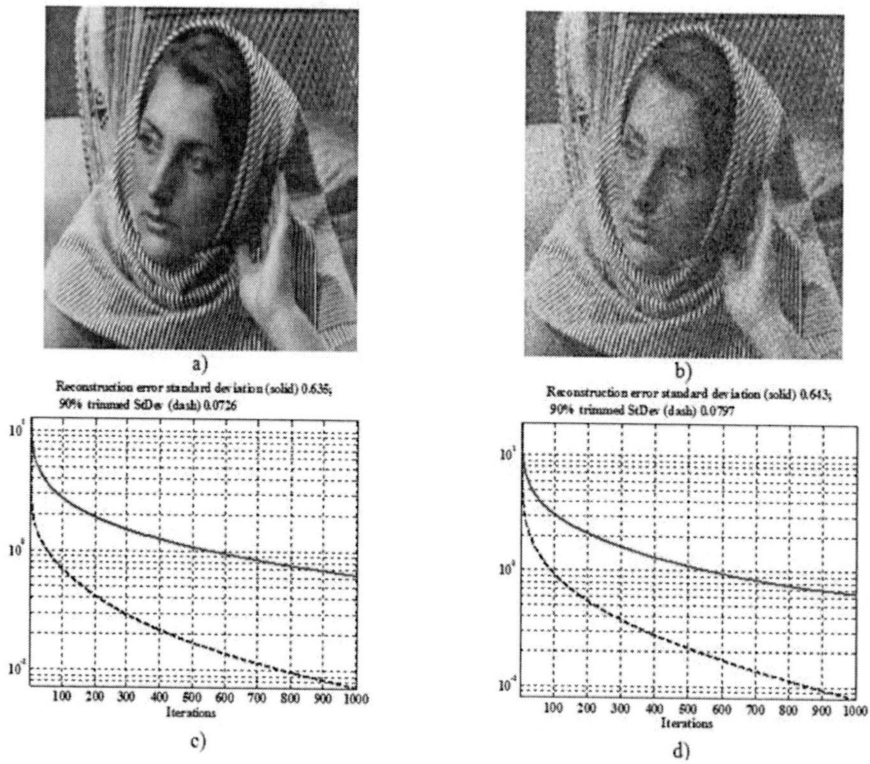

Figure 11. a), b) - images reconstructed from a sampled noiseless test image and, correspondingly, from the same sampled image contaminated with additive uncorrelated Gaussian noise with standard deviation 20 gray levels; c), d) - corresponding graphs of RMS of all (solid lines) and of the smallest 90% (dash lines) of reconstruction errors versus the number of iterations.

Image Antialiasing Pre-Filtering

As it was indicated in the previous section, image pre-filtering before sampling is necessary in order to avoid aliasing artifacts and to secure convergence of the iterative algorithm to an image with bounded spectrum. In ordinary imaging devices, antialiasing pre-filtering is carried out by imaging optics together with apertures of image photo sensors. ASBSR-method envisages, generally, choosing an EC-zone shape for every particular image. Ordinary photo sensors are not capable of implementing this choice. As a solution to this problem, usage of multiple aperture sensors, such as those suggested in Ref. [17], can be proposed. The required effective aperture of these sensors defined by inverse Fourier transform of the chosen EC-zone shape is synthesized by combining, with appropriate weights, signals from individual apertures.

The use of conventional single aperture sensors is also possible at the expense of a certain worsening of accuracy of approximating images by their BS-copies, reconstructed by the ASBSR-method. As such a universal ("all-purpose") EC-zone shape, a "pie-sector" shape can be suggested, which fits majority of natural images quite well. As it was mentioned in Sect. 4, substantial variations of EC-zone shape parameters do not lead in practice to substantial variations of image approximation RMSE. This is why the worsening image approximation accuracy due to the use an "all-purpose" shape instead of a "dedicated" one will be, as a rule, not dramatic. For instance, total RMS error 4.72 of reconstruction of test image "Blood Vessels 512" using an "all-purpose" pie-sector zone (Figure 12) is only 7.7% larger than reconstruction RMSE 4.38 obtained, when a triangular EC-zone that fits image spectrum (Figure 9) was used.

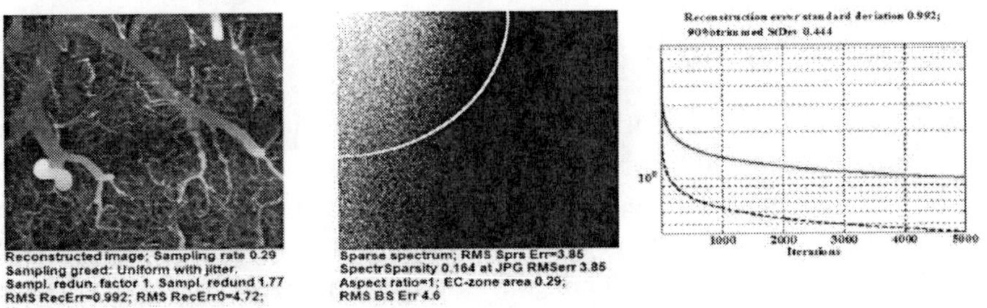

Figure 12. An example of reconstruction of test image "Blood Vessels 512" using a pie-sector EC-zone of the same area as that of the triangular shape (Figure 9), which better fits the image sparse spectrum.

Computational Complexity

The computational complexity of the reconstruction algorithm per iteration is determined by the complexity *O(2NlogN)* of floating point operations for direct and inverse fast transforms plus *O(N)* replacement operations for sample-wise modifications of data in image domain (*M* operations) and in its transform domain (*N* - *M* operations). The order of magnitude of time required for one iteration can be estimated from these data: elapsed time for Matlab direct or inverse DCT of an array of $N = 512 \times 512$ numbers implemented on a PC "Lenovo ThinkPad X201" with processor Intel i7 and operating system Windows-7 is 52 msec.

7. Other Possible Applications of the ASBSR-Method of Image Sampling and Reconstruction

The above discussed task of reconstruction of images of N samples from $M < N$ sampled data can be considered as a special case of under-determined inverse imaging problems. One can expect that the found solution to this problem, the bounded spectrum (BS-) image reconstruction, may find application for solving other under-determined inverse imaging problems as well. We illustrate this possibility in five applications: (i) demosaicing color images, (ii) in-painting of occlusions in images, (iii) image reconstruction from their sparsely sampled or decimated projections, (iv) image reconstruction from their sparsely sampled Fourier spectra, and (v) image reconstruction from the modulus of its Fourier spectrum.

Demosaicing Color Images

Individual cells of light sensitive arrays of modern color cameras are split into three groups allocated for corresponding Red, Green and Blue components of color images. It is made by means of placing, in front of cells, arrays of red, green and blue filters. Most frequently used is Bayer's arrangement of filters ([18]) shown in Figure 13, a). We will call this arrangement "Regular Bayer". There is also an alternative method shown in Figure 13, b). It assumes the regular Bayer arrangement of green pixels and pseudo-random arrangements of red and blue pixels ([19]). We will call this arrangement "Semi-random". In both arrangements, density of green pixels is 0.5, and densities of red and blue pixels are correspondingly 0.25 of the overall pixel density, i.e., half of all cells of the sensor array are allocated for green pixels, one quarter is allocated for red, and one quarter for blue pixels.

Images captured by color cameras are displayed in full resolution of the sensor, for which purpose half of pixels of green image components and three quarters of pixels of red and blue image components should be reconstructed from their corresponding available pixels. This reconstruction of unavailable color separated pixels is called image demosaicing. As a demosaicing mechanism, different interpolation methods are discussed in literature ([19]). In practice, bilinear interpolation is ordinarily used. We will call this method of color image demosaicing "Bilinear demosaicing".

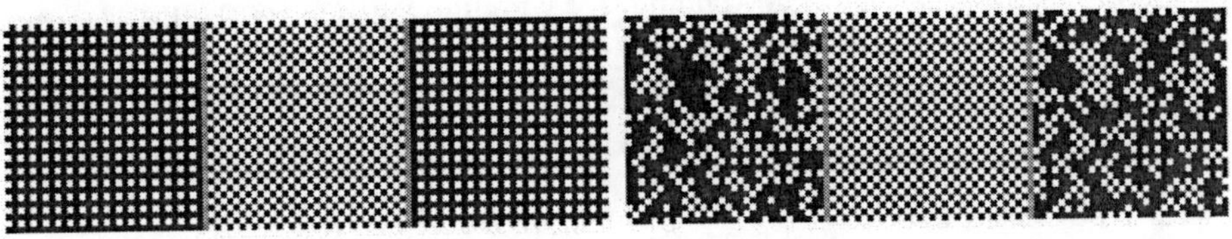

a) b)

Figure 13. Regular Bayer (a) and semi-random (b) arrangements of color separated pixels in photosensitive arrays of modern color cameras (from left to right: R, G, and B components).

Bilinear interpolation is one of the simplest 2D interpolation methods. It, however, is far from being perfect since it distorts images in their baseband and introduces spectrum aliasing errors. From the theory outlined in Sect. 2, it follows that, given the number of available pixels, optimal image reconstruction in terms of reconstruction RMSE is the Bounded Spectrum (BS-) reconstruction, i.e., reconstruction of images with spectrum bounded by chosen for them EC-zone shapes. In application to image demosaicing, the most natural is using as the spectrum bounding shape the "all-purpose" shape in form of a pie-sector that allows implementing image anti-aliasing pre-filtering before their sampling by camera optics and by apertures of cells of its photo sensitive arrays. We will call this method of demosaicing "BS-demosaicing" performance of image BS-demosaicing in comparison with that of Bilinear demosaicing is illustrated in Figure 14 on a test image Light House 512 for two cases: regular Bayer arrangement and Semi-random arrangement of pixels, correspondingly.

In the top row in Figure 14 shown are (from left to right) a test color image of 512x512 pixels and sparse spectra (white dots) of the image red, green and blue components that contain spectral coefficients, which reconstruct corresponding images with reconstruction RMSE equal to that of JPEG image coding. Solid lines in the spectra images indicate borders of EC-zones used for image reconstruction by the BS-method. According to the rates of pixels in red, green, and blue image components, relative areas of the EC-zones are set to 0.25, 0.5, and 0.25, correspondingly. In the middle and bottom rows of the figure shown are bottom right 128x128 pixel fragments of the test and of demosaiced images obtained by the BS- and by Bilinear demosaicing methods for regular Bayer (middle row) and semi-random (bottom row) color pixel arrangements. The fragments are magnified for better visibility of demosaicing artifacts. One can easily see in these images that the Bilinear demosaicing produces much more color artifacts compared to those of the BS-demosaicing, for which the artifacts are quite subtle.

Table 2. RMS reconstruction errors for the Bilinear and BS-demosaicing

"Regular Bayes" color pixel arrangement				
BS-demosaicing, RMS of reconstruction errors	Red component 9.11	Green component 4.38	Blue component 9.43	Total average 7.98
Bilinear demosaicing RMS of reconstruction errors	Red component 11	Green component 6.31	Blue component 11.2	Total average 9.76
"Semi-random" color pixel arrangement				
BS-demosaicing, RMS of reconstruction errors	Red component 9.55	Green component 4.36	Blue component 9.64	Total average 8.23
Bilinear demosaicing RMS of reconstruction errors	Red component 13	Green component 6.26	Blue component 12.6	Total average 11.05

Numerical data on reconstruction RMSE for both demosaicing methods and Regular Bayer and Semi-random pixel arrangements are given in Table 2. According to them, BS-demosaicing outperforms the Bilinear Bayer demosaicing in terms of reconstruction RMSE as well. In particular, total reconstruction RMSE for the BS- and Bilinear methods are, correspondingly, 7.98 and 9.76 for the case of "Regular Bayes" arrangement and 8.23 and 11.05 for "Semi-random" arrangement of color pixels. Relatively large values of image reconstruction errors can be attributed to the fact that, as one can see from image spectra, quite large portions of spectra red, green, and, especially, of blue components fall outside the EC-zones used for image reconstruction.

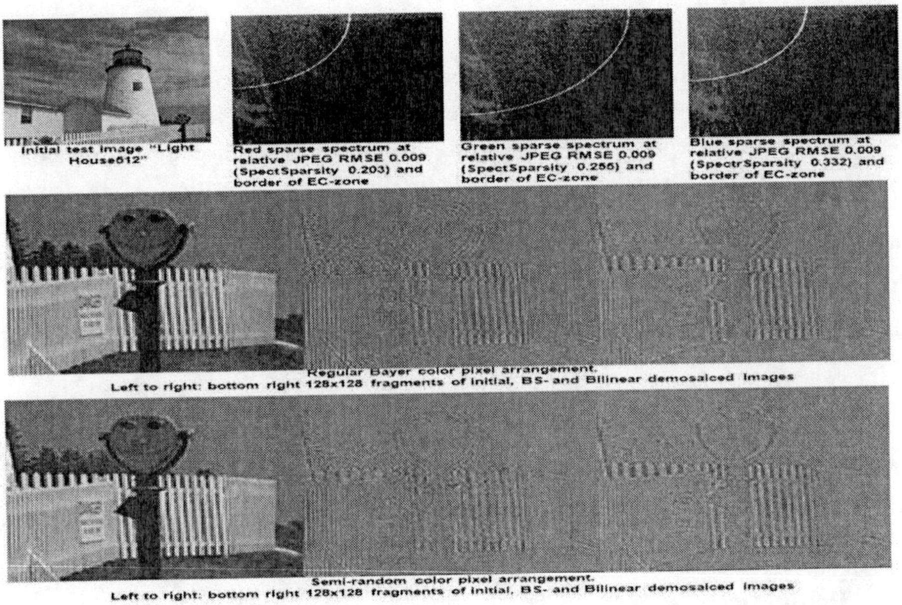

Figure 14. Comparison of BS-demosaicing and Bilinear demosaicing of a test color image with regular Bayer and semi-random pixel arrangements. Top row (left to right): test image of 512x512 pixels and sparse spectra of its red, green, and blue components (white dots) along with borders of the corresponding chosen EC-zones (white solid line). Middle and bottom rows (left to right): 128x128 pixel bottom right fragments of the initial test image and of demosaicing errors of BS- and Bilinear demosaiced images for "Regular Bayer" and "Semi-random" color pixel arrangements, correspondingly.

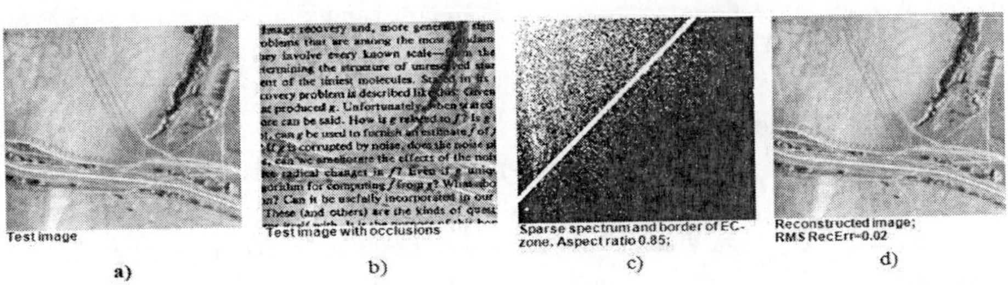

Figure 15. An example of using image BS-reconstruction for image in-painting: a) test image; b) test image occluded with a text; c) test image sparse DCT spectrum at JPEG RMSE4.37 (white dots) and border of spectrum EC-zone used for image reconstruction; d) reconstructed test image (reconstruction RMSE 0.02 gray levels; peak signal to RMSE ratio 82.1 dB).

Image in-Painting

Image in-painting is recovering images obstructed by foreign occlusions. In frequent cases image foreign occlusions can be reliably detected by means of one or another image segmentation method. In such cases, in-painting of image occlusions can be carried out with exactly the same algorithm as the above described iterative algorithm of image BS-reconstruction from sparse samples, in this case from those samples that are found not occluded. An illustrative example is presented in Figure 15.

In this example a test image (Figure 15, a) is occluded by a text (Figure 15, b), which is completely black and thanks to this its pixels can be easily detected by their gray level. Figure 15, d) presents the image BS-reconstructed from its samples not marked as occlusions. The image EC-zone used for the reconstruction is shown in Figure 15, c).

Image Reconstruction from Their Sparsely Sampled or Decimated Projections

In computed tomography, it quite frequently happens that body slice occupies only a fraction of the image entire area. This means that slice projections are Radon transform "Bounded Spectrum" functions. Therefore, whatever number of projections or of their samples is available, a certain number of additional projections or samples, commensurable, according to the discrete sampling theorem ([14], [15]), with the size of the slice empty zone, can be obtained, and the corresponding resolution increase in the reconstructed images can be achieved using the method of image BS-reconstruction ([14]). This option is illustrated in Figure 16 and Figure 17.

Figure 16 illustrates recovery of missing samples of an image slice projections sampled in random positions. In the experiment, it was found by means of simple segmentation of the test image shown in Figure 16, a) that the outer 55% of the image area is empty. Then the same percentage of projection samples selected randomly using the Matlab random number generator were zeroed (Figure 16, c). The rest of samples was used for recovering missing samples and, correspondingly, for image reconstruction by means of the iterative reconstruction algorithm identical to the above described image BS-reconstruction algorithm except that direct and inverse Discrete Cosine Transforms were replaced by direct and inverse Discrete Radon Transforms. At each iteration of the algorithm, the current set of projections is subjected to inverse Radon transform for obtaining a current estimate of the reconstructed image. Then the outer empty area of the reconstructed image is zeroed and this corrected image is subjected to direct Radon transform for obtaining a next estimate of slice projections. In the obtained projections, their available samples are restored and the process repeats.

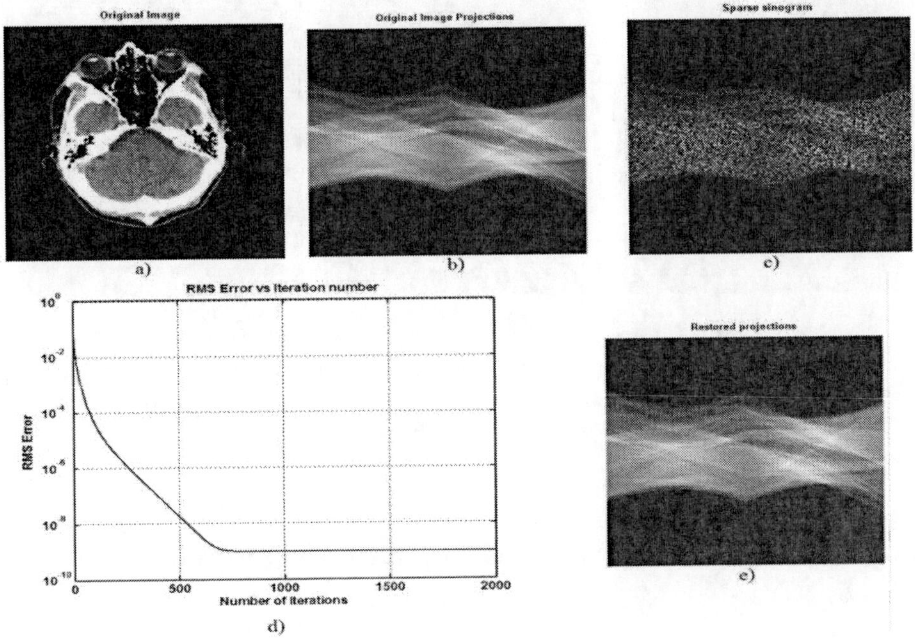

Figure 16. Recovery of randomly sampled slice projections: a) test image; b) its slice projections; c) slice projections randomly sub-sampled with rate 0.45; d) plot of slice projections reconstruction RMSE versus the iteration number; e) the recovered slice projections.

The plot of RMS of slice projections reconstruction errors versus the iteration number shown in Figure 16, d) and the result of recovering missing samples (Figure 16, e) show that virtually perfect recovery of missing 55% samples of slice projections is possible with the iterative reconstruction algorithm after few hundreds of iterations.

Figure 17 illustrates that recovery of completely missing projections is also possible. In the experiment, every second projection of the test image shown in Figure 16, a) was removed (Figure 17, b) and then all initial projections were recovered (Figure 17, c) by the above described iterative algorithm that makes use of the fact that the outer 55% part of the image area is known to be empty. This result allows suggesting that for such cases, when half or bigger part of the image area is known to be empty, one can use the method of image BS-reconstruction to achieve image reconstruction with super-resolution that corresponds to larger number of image projections than are available.

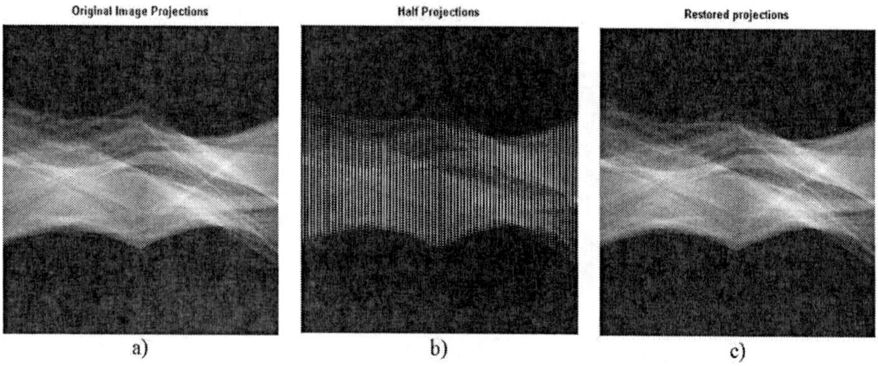

Figure 17. Recovery of missing image projections: a) original projections of the test image of Figure 16, a); b) decimated projections with every second projection removed; c) slice projections recovered from decimated slice projections (b) using the iterative reconstruction algorithm.

Image Reconstruction from Their Sparsely Sampled Fourier Spectra

There exist some imaging devices (e.g., some healthcare scanners), where sampling is done in a transform domain. The proposed ASBSR-method can be used in such devices in those frequent cases, when it is known that object image is surrounded by some empty space. Figure 18 demonstrates this option on an example of image reconstruction from its sparsely sampled Fourier spectrum.

In this example, Fourier spectrum of a test image bounded by a circular binary image mask was randomly sampled with sampling rate equal to the ratio of the image bounding circle area to the area of the entire image frame. Additionally, the spectrum was bounded by a circular binary spectral mask with radius equal to the highest horizontal and vertical spatial frequency of the baseband. This gives an additional $1-\pi/4=21.5\%$ saving in the number of spectrum samples.

For image reconstruction, an iterative algorithm was used. At each iteration, the iterated spectrum is inverse Fourier transformed for obtaining an iterated reconstructed image and then the latter is multiplied by the bounding circular image mask and Fourier transformed. Samples of the obtained spectrum in positions of available original ones are restored, and the spectrum is bounded by the circular binary spectral mask to form an iterated spectrum for the next iteration. As a spectrum zero order approximation, from which the iterative reconstruction starts, the initial sparsely sampled and bounded spectrum was used.

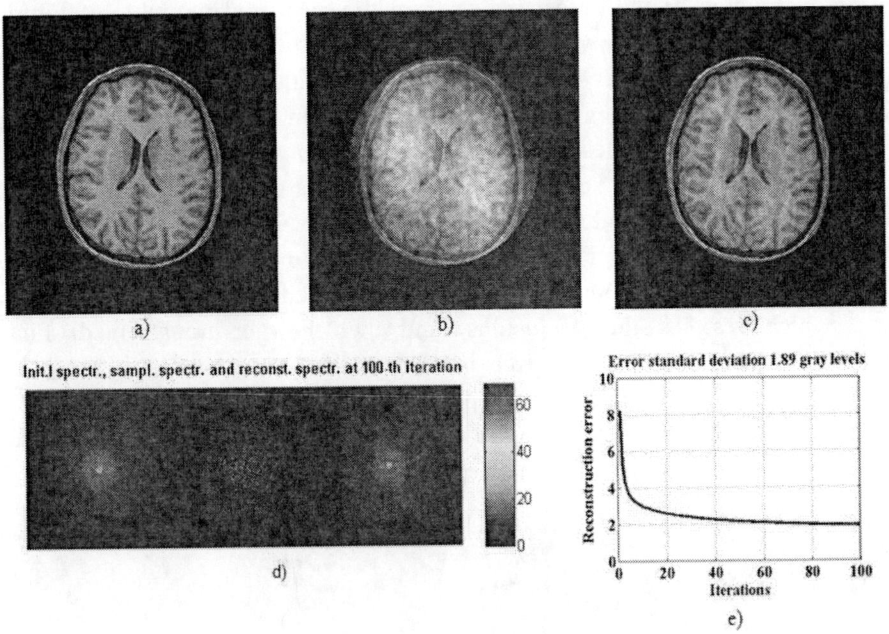

Figure 18. Image reconstruction from its sparsely sampled spectrum: a)test image bounded by a binary circular mask with radius equal to 0.35 of the image size; b) reconstructed image at the first iteration, in which the circular bounding mask can be seen; c) reconstructed image at the 100-th iteration (RMS reconstruction errors 1.89,peak signal to RMSE ratio is 49 dB); d) from left to right: Fourier spectrum of the test image, this spectrum randomly sampled with sampling rate $\pi \text{ x } 0.35^2 \text{ x} - \pi/4 = 0.3$, and reconstructed spectrum (for display purposes, absolute values of spectral samples are displayed raised to power 0.3 and are shown color coded with Matlab color map “jet”, color on-line); e) plot of reconstruction RMSE vs the number of iterations.

Image Reconstruction from the Modulus of Its Fourier Spectrum

In order to enable image reconstruction from the modulus of its Fourier spectrum using the suggested ASBSR-method, object should be imaged through a randomized binary (opaque-transparent) mask that produces occlusions in the object image. Fraction of the transparent area of the mask should be equal to or larger than the required by the ASBSR-method sampling rate, i.e., than the fraction of the area occupied in the spectrum baseband by the chosen for this image EC-zone of the image DCT spectrum. Measured is the modulus of Fourier spectrum of the object occluded by the mask.

Reconstruction of the object image from the measured modulus of Fourier spectrum of its occluded copy is carried out in two stages. At the first stage, object image with occlusions is reconstructed from the modulus of its Fourier spectrum using an iterative Gerchberg-Papoulis type algorithm. Iterations reconstruct the image with occlusions and estimates of phase component of its Fourier spectrum. At each iteration, a current spectrum phase component estimate is combined with the measured spectrum modulus to form a complete spectrum estimate, which is inverse Fourier transformed to obtain a current estimate of the reconstructed image with occlusions. Then this image is multiplied by the binary occlusion mask, which restores the occlusions, and then Fourier transformed. Phase component of the obtained spectrum is used as the next estimate of the image phase component, and iterations are repeated. As a starting zero order estimate of the image spectrum phase component, phase component of the Fourier spectrum of the binary occlusion mask can be used.

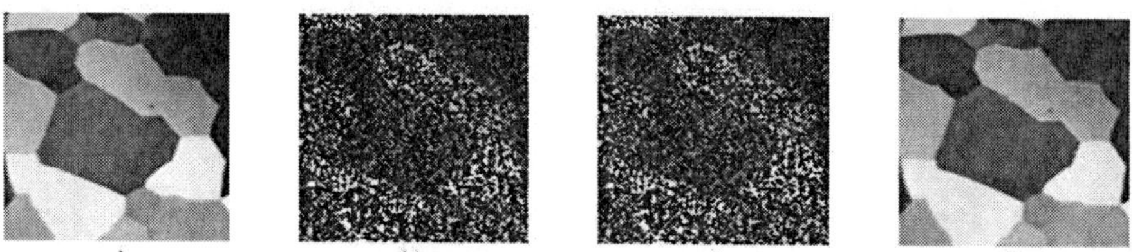

a) b) c) d)

Figure 19. Image BS-reconstruction from the modulus of its Fourier spectrum: a) test image; b) test image occluded by randomly placed 3x3 pixel opaque squares; c) occluded image reconstructed from the modulus of its Fourier spectrum (reconstruction RMSE is 10^{-3}; peak signal to RMSE ratio 107 dB); d) test image BS-reconstructed from the reconstructed occluded image (c) (reconstruction RMSE is 1.79; peak signal to RMSE ratio is 43.1 dB).

At the second stage of image reconstruction, the reconstructed image with occlusions is used for BS-reconstruction of the entire image in the same way as in the above described image in-painting. An illustrative example is shown in Figure 19.

Conclusion

The problem of minimization of the number of measurements needed for digital image acquisition and reconstruction with a given accuracy is addressed. Specifically:

- It is shown that the theoretical lower bound of sampling rate sufficient for reconstructing an image of N samples with a given RMSE is defined by image spectrum sparsity for this level of image sparse approximation error, i.e., by the ratio of the area of image spectrum Energy Compaction (EC-) zone formed from the largest spectral coefficients, which reconstruct the image with the given RMSE, to the area of the image baseband defined by its N samples.
- Data saving capability of compressed sensing in image acquisition is disputed and it is shown that, for the range of sparsities of natural images from 10^{-1} to 10^{-3}, the required number of measurements must exceed the theoretical minimum from 3 to 8 times, which implies that compressed sensing is far from reaching the theoretical minimum of signal sampling rate.
- A method of image Arbitrary Sampling and Bounded Spectrum Reconstruction (ASBSR-method) is described capable of reconstructing images sampled with rates that come near to the theoretical minimum. The method assumes representing images in a domain of one of transforms with sufficiently good energy compaction capability, approximating image spectra EC-zones by one of a few standard shapes, image sampling in arbitrary positions, and reconstructing images with spectrum bounded by the chosen shape. The method is insensitive to noise in sampled data and allows to secure obtaining a desired resolving power of reconstructed images.
- The workability of the method is demonstrated by results of its experimental verification using DCT as the image transform, image sampling over uniform sampling grids with random jitter, and an iterative Gerchberg-Papoulis type reconstruction algorithm. These results confirm that the sampling rate required by the method is equal to the theoretical minimum for the chosen EC zone shape, i.e., to the shape's relative area. This means that, for the class of images with the chosen EC-zone, the

method is non-redundant. Sampling rates of particular images tested in the experiments exceeded their actual sparsities 1.5-1.8 times. This is the order of magnitude of the ultimate sampling redundancy of the method associated with approximating image spectra EC-zones by one of the standard shapes.

- Examples of application of the ASBSR-method for color image demosaicing, image in-painting, image reconstruction from sparsely sampled or decimated projections, image reconstruction from its sparsely sampled Fourier spectrum, and image reconstruction from the modulus of its Fourier spectrum demonstrate potential applicability of the method for solving other underdetermined inverse imaging problems.

References

[1] Donoho D., *IEEE Trans. Inform. Theory* 52(4), 2006, pp.1289–1306.

[2] Candès, E., *Proc. Int. Congress of Math.*, Madrid, Spain (2006).

[3] Candès E., Romberg J., and Tao T., *Commun. Pure Appl. Math.* 59 (8), 2006, 1207-1223.

[4] Bucholtz F. and Nichols J. M., *Opt. Photonics News* 25, 2014, pp. 44–49.

[5] Michaud S, *Opt. Photonics News*, Nov. 2014, pp. 21 -23.

[6] Candès E. and Wakin M., *IEEE Signal Processing Magazine*, 25(2), March 2008, pp. 21 – 30.

[7] Willett R., Marcia R., and Nichols., *Opt. Eng.* 50(7), July 06, 201, pp.1 -13.

[8] Nichols J. M. and Bucholtz F., *Opt. Express* Vol. 19, No. 8, 2011, pp.7339-7348.

[9] Kotelnikov V. A., *Material for the First All-Union Conference on Questions of Communication, Izd. Red. Upr. Svyazi RKKA*, Moscow, 1933 (in Russian. English translation: http://ict.open.ac.uk/classics/ 1.pdf).

[10] Shannon C., *Bell Sys. Techn. Journ.,* vol. 27,1948, No. 3, 379-423, No. 4, 623-656.

[11] Donoho D. L. and Tanner J., *IEEE Trans. Inf. Theory* 56, Issue 4, April 2010 2010, 2002 -2016.

[12] Yaroslavsky L. P., *Opt. Eng.,* 54(7), 2015, pp. 079701-1-079701-4.

[13] Yaroslavsky L. P., *Journal of Optics.* v. 19, No 5, May 2017, 055706.

[14] Yaroslavsky L., Shabat G., Salomon B. G., Ideses I. A., and Fishbain B., *J. Opt. Soc. Am. A* 26, 2009, pp.566–575.

[15] Yaroslavsky L., *Theoretical Foundations of Digital Imaging*, CRC Press, Roca Baton.

[16] Yaroslavsky, L., *Digital Holography and Digital Image Processing*, Kluwer Academic Publishers, Boston, 2004 London, N.Y., 2013).

[17] Fiete R. D., Multiple aperture imaging system, US patent US 6,943,946 B2, Sep. 13, 2005 (https://www.google.com/patents/US6943946).

[18] Bayer, B., "Color imaging array", U.S. Patent 3,971,065, Eastman Kodak Company, July 20, 1976.

[19] Zapryanov G., Nikolova I., In: *Fourth International Bulgarian-Greece Conference Computer Science 08*, 2008, 133. (http://www.utp.bg/wp-content/uploads/2016/01/11.pdf).

In: Compressed Sensing
Editor: J. M. Sheppard

ISBN: 978-1-53613-082-9

Chapter 2

THE COMPRESSED SENSING SUPER RESOLUTION OF MULTIPLE IMAGES

Wael H. Saafin, PhD*
Department of Computer Science,
Loughborough University, Loughborough, UK

Abstract

Compressed Sensing (CS) is a new technology that simultaneously acquires and compresses images reducing acquisition time and memory requirements to process or transmit them. It establishes that a sparsely representable image/signal can be recovered from a highly incomplete set of measurements or projections of the image. Image Super Resolution (SR) is an important post-processing technique where multiple input images are super resolved to obtain one or more images of higher resolution and better quality. SR necessitates a good image registration procedure in order to obtain a High Resolution (HR) image of enhanced quality. Such quality should overcome image degradation due to system hardware, optical and spatial limitations.

In this chapter we study a novel framework to obtain HR images from CS imaging systems capturing multiple Low Resolution (LR) images of the same scene. The assumption that when an image admits a sparse representation in a transformed domain, a blurred version of it will also be sparse in the transformed domain, allows us to recover blurred images from CS observations. Similarly, a warped, blurred, and down-sampled

*E-mail address: waelsaafin@hotmail.com.

LR image is expected to be also sparse in a transformed domain and hence can be reconstructed from the corresponding CS observation. The proposed Compressed Sensing Super Resolution (CSSR) approach, combines existing CS reconstruction algorithms with an LR to HR approach based on the use of a new robust sparsity promoting prior based on super Gaussian regularisation.

The CSSR problem has multiple unknowns, mainly the HR image, the transformed coefficients corresponding to each LR image, and the motion vectors. The unknowns can be estimated sequentially or simultaneously. The method we study in this work is a joint estimation framework where LR reconstructions and HR estimation are carried out simultaneously by alternating between them including the automatic estimation of registration parameters. The advantages of the joint estimation over the sequential approach and the advantage of utilising multiple input images over a single one are also discussed.

The CSSR approach converts the constrained joint optimisation problem into a sequence of unconstrained sub-problems using Alternate Direction Method of Multipliers. The HR image estimation sub-problem is solved using majorisation-minimisation. The registration parameters are estimated using two different approaches, which are compared in this work. The CS reconstruction sub-problem becomes an ℓ_1 minimisation subject to a quadratic constraint.

The proposed framework has been tested on grayscale simulated and synthetically compressed real natural images, which were compressed using a measurement matrix that can be synthesised practically. This is compared with state-of-the-art SR algorithms which do not use CS in their work, and with state-of-the-art image deconvolution algorithms from CS images, which do not use down-sampling in their work. The proposed CSSR performs favourably compared to them. Moreover, the proposed CSSR has been tested when both compression and down-sampling are present.

Two important CSSR applications are investigated: Passive Millimeter Wave (PMMW) images and colour images. The performed CSSR experiments on synthetically compressed real millimeter wave and colour images demonstrate the capability of the CSSR framework to provide very good quality super resolved images from multiple low resolution compressed acquisitions. PMMW imaging is used for threat detection and the CSSR is expected to improve threat detection rates. For CS colour images consisting of Red, Green, and Blue (RGB) channels which were sensed separately, the registration parameters are jointly estimated from the three channels simultaneously using the CSSR framework.

Keywords: Energy of vacuum state, compressed sensing, image super-resolution, image reconstruction, image registration, inverse problems, millimeter wave images, colour images

PACS: 05.45-a, 52.35.Mw, 96.50.Fm

AMS Subject Classification: 53D, 37C, 65P

This work is devoted to estimate High Resolution (HR) images from multiple Low Resolution (LR) observations acquired using Compressed Sensing (CS) techniques. The proposed framework combines Super Resolution (SR) reconstruction with CS acquisition.

CS imaging lowers the sampling frequency beyond that required by the Nyquist-Shannon theorem, and hence lowers transmission rate and bandwidth requirements. The smaller data size obtained during the acquisition process lowers the memory requirements to save or transmit. Another important benefit CS offers is the lower acquisition time, due to the smaller data size, which is vital in applications like medical imaging and millimeter wave images.

Furthermore, SR offers a higher pixel density and overcomes the limited resolution imaging. The higher resolution enhances the image quality by showing more details of the original scene. Resolution enhancement is needed in computer vision applications like pattern recognition and analysis of images, in medical imaging for diagnosis, and in applications that require specific areas of the image to be zoomed, like surveillance cameras, satellite applications, and threat detection devices. Moreover, SR promises a better use of high performance screens, like High Definition Liquid Crystal Displays (HD LCDs).

In the Compressed Sensing Super Resolution (CSSR) framework classical SR methods are combined with existing CS methods. The experimental results to be presented will prove the applicability of this framework, and the good quality of the output images. A background is presented next.

Independently of how well designed an imaging system is, its performance comes to a certain limit. Limitations on image spatial resolution lead to the use of SR techniques to improve the pictorial perception of the human eye and/or automatic machines. After a certain limit, an imaging system suffers from many hardware problems, such as limited pixel density, lens point spread function and aberration effects. Other problems may accompany the acquisition process, such as blurring due to motion of the camera or objects in the scene. The acquired images can be considered as decimated LR versions of the original scene.

These LR images can be used to estimate an HR image using SR techniques.

SR techniques are post-processing techniques, they use multiple LR images as input images to estimate one, or more, HR output image(s). The process increases the spatial resolution, removes degradations in the input images, and increases the high frequency content of the scene image.

Traditionally, all pixel values of the LR image have to be acquired. However, this information can be successfully recovered from only a small number of linear projections utilising CS techniques. This work studies how to estimate SR images from multiple CS observations.

1. Super Resolution

As we have already indicated, both hardware limitations, such as sensor size, camera speed, cost, and optical limitations such as blurring action, etc. make it necessary to explore alternative ways to enhance the resolution of an image.

Instead of decreasing the effects of these limitations by working on the camera itself, resolution enhancement or SR techniques processes the images acquired under these limitations and try to reconstruct one or more HR image(s). This post-processing step utilises multiple input LR images of the same scene to estimate images of higher spatial resolution, with less degradation effects, and better quality. The estimated image should not only provide a better visualisation (visual quality issue), but also extracts additional information details from the input images (recognition issue).

The multiple input images can be acquired successively by the same camera, or by multiple cameras imaging the scene simultaneously. In both cases, it is expected to result in some shifts between the various acquired images. The shifts may be due to a small camera movement with respect to the scene, for example. These subpixel shifts or displacements give the added information that enables to efficiently estimate the HR image.

The acquired images can be considered down-sampled, blurred, and warped versions of the original HR image. Assume that we have the LR image sequence $\{\mathbf{y}_q\}$ where $q \in \{1, \cdots, Q\}$. Each $\mathbf{y}_q$ is represented by a $D \times 1$ vector. The observation process of the LR image sequence can be modelled as follows,

$$\mathbf{y}_q = \mathbf{A}\mathbf{H}_q\mathbf{C}(\mathbf{s}_q)\mathbf{x} + \mathbf{n}_q = \mathbf{B}_q(\mathbf{s}_q)\mathbf{x} + \mathbf{n}_q, \text{ for } q = 1, \ldots, Q, \tag{1}$$

where $\mathbf{A}$ is a $D \times N$ down-sampling matrix, $\mathbf{H}_q$ is an $N \times N$ blurring matrix, $\mathbf{C}(\mathbf{s}_q)$ is the $N \times N$ warping matrix corresponding to a 3×1 motion vector $\mathbf{s}_q$,

$\mathbf{x}$ is the $N \times 1$ HR image vector, and $\mathbf{n}_q$ is a $D \times 1$ vector representing the noise assumed to be additive. Elements of the motion vector $\mathbf{s}_q = [\theta, c, d]^t$ correspond respectively to rotation angle, horizontal and vertical displacements. $\mathbf{B}_q(\mathbf{s}_q)$ is a $D \times N$ matrix modelling the acquisition system. The zooming factor is defined to be $P \geq 1$ and represented by the factor of increase in each dimension of the image, hence the factor of increase in resolution is $P^2 = N/D$. Figure 1 shows an example of application of the four degradation steps of this acquisition model to the Cameraman image.

Figure 1. Degradation process illustrative example. (a) Original Cameraman image, (b) Warped image, $\theta = -0.1047rad$, $c = 3$, $d = 2$, (c) Blurred image, Gaussian blur with variance 3, (d) Down-sampled image, P=4, (e) Noised image, white Gaussian noise SNR=40dB.

Warping, in our context, is the displacement affecting the acquired images with respect to the reference image or the original image. This displacement can be horizontal or vertical when all the objects in a scene are shifted in the $x-$, $y-$ direction by the same distance c, d respectively. It may also be rotational, when the observation is a rotated version of the reference image, by a rotational angle

θ. The three parameters (θ, c, d) form the motion vector of the observation. An example of a warped image is shown in Figure 1(b), where the motion vector used was $[-0.1047, 3, 2]^t$, θ is in radians, and the last two entries are the number of pixel shifts in the horizontal and vertical directions respectively.

A blurred image may be the result of many factors affecting the imaging system, like motion blur and out-of-focus blur, among others. An object moving within a scene or the motion of the camera itself may result in motion blur. Out-of-focus blur depends on many parameters like focal length, camera aperture size and shape, distance between the camera and the observed scene, wavelength of the incoming radiation, and the effects due to diffraction [Amizic et al., 2013a]. However, in this work we are not interested in the type of blur or the phenomena affecting it. Instead, we include an approximation that models its effect on the observed image. In Figure 1(c) a Gaussian blur, with variance 3, has been added to the warped image.

Down-sampling, or decimation, is the process of reducing the sampling rate of an image signal. Let us describe the down-sampling of the image shown in Figure 1(c), to obtain the image in Figure 1(d). In the two dimensional representation of the Figure 1(c) image, divide the whole into square blocks, each of $P \times P$ size. Then from the P^2 pixels in each block, keep one (or the mean) pixel and discard all the others. The down-sampled image is represented only by the pixels kept, and each dimension of the original image will be P times that of the decimated version. In vector form, the length of the resulting image will be N/P^2. In Figure 1(d), the zooming factor in each dimension is $P = 4$, hence the block size is 4×4. The pixels kept form the down-sampled image shown in Figure 1(d). Notice that the factor of increase in resolution is $P^2 = 16$. Finally, Figure 1(e) shows a noised version of the down-sampled image after adding white Gaussian noise with SNR=40 dB.

SR reconstruction collects information from various LR images to estimate one HR image. Figure 2 illustrates the process, for a zooming factor P=2, and shows how the information from each image is shared in the estimation process of the HR image pixels. The crossings made by red dashed lines in Figure 2 represent the new added pixels to the estimated HR image. Notice that the HR image is not in its final stage, and still there are some works to be applied to it, as it will be clear later in this chapter. SR reconstruction starts from the LR image sequence represented in the upper row of Figure 2.

SR research started in 1984 with the pioneer work by Tsai and Huang [Tsai and Huang, 1984]. Many SR methods have been proposed, which can be

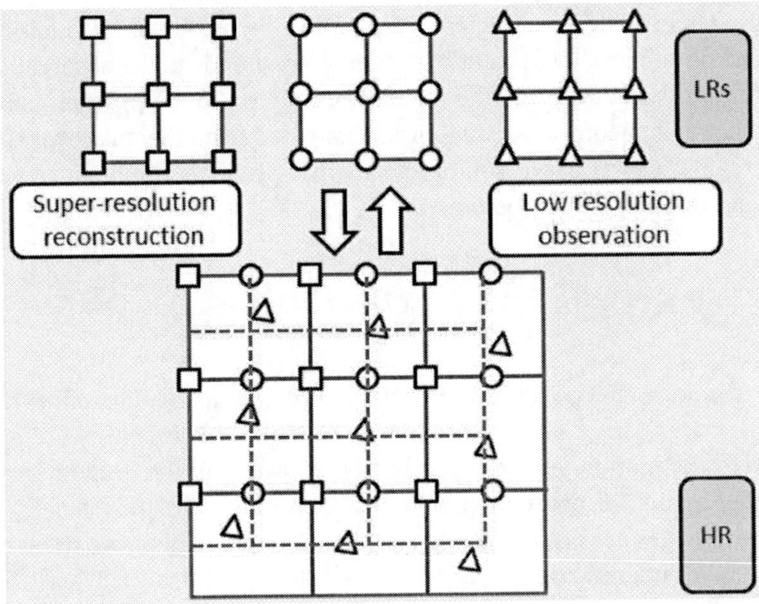

Figure 2. SR Estimation. Upper part: three LR images each of resolution= 9. Lower part: estimated HR image of resolution=36, formed by the original pixels in LR observation and the crossings made by the red dashed lines (P=2).

grouped into the following classes: Frequency domain based approach [Tsai and Huang, 1984, Ji and Fermuller, 2009, El-Khamy et al., 2005], interpolation based approach [Bose et al., 2006], Bayesian approach [Tipping and Bishop, 2006, Villena et al., 2009, Schultz and Stevenson, 1996, Villena et al., 2013, Pickup et al., 2009, Belekos et al., 2010, Babacan et al., 2011], regularisation based approach [Farsiu et al., 2004, Zibetti et al., 2011, Marquina and Osher, 2008, He et al., 2007, Panda et al., 2015], and learning based approach [Yang et al., 2010b, Kim and Kwon, 2010, Wang et al., 2010]. There are some variants of the SR problem: an HR image can be reconstructed from multiple LR observed images, it is also possible to reconstruct an HR image sequence from an observed LR image sequence, and finally an HR image can be reconstructed from only one LR observed image using the learning based approach.

This work belongs to the regularisation-based approach. The basic idea of this approach to solve the ill-posed inverse SR problem, is to introduce in the cost function to be optimised a regularisation term constraining the estimation process. This should be sufficient to solve for the unknown HR image. From a Bayesian point view, the data fidelity term corresponds to the observation model probability distribution, and the regularisation term to the prior probability distribution which incorporates our a priori knowledge on the unknown HR image.

A regularisation based solution to the SR problem can be achieved by the following minimisation problem,

$$\hat{\mathbf{x}}, \hat{\mathbf{s}} = \arg\min_{\mathbf{x},\mathbf{s}} \frac{\beta}{2} \sum_{q=1}^{Q} \| \mathbf{B}_q(\mathbf{s}_q)\mathbf{x} - \mathbf{y}_q \|^2 + \gamma\, \mathbf{Q}(\mathbf{x})\,, \tag{2}$$

where β and γ are positive parameters, $\mathbf{Q}(\mathbf{x})$ is a regularisation term, and $\mathbf{s} = (\mathbf{s}_1, \cdots, \mathbf{s}_Q)$ is a set of motion vectors. If the LR images were previously registered the motion vectors $\mathbf{s}_q$ would be known, and the only unknown to be estimated by the SR process would be $\mathbf{x}$. Otherwise, which is usually the case, the input images are not registered and the motion vectors have to be estimated like the other unknowns.

Next we present some regularisation terms, $\mathbf{Q}(\cdot)$, utilised in the SR literature. For some of these regularisers, we describe the corresponding SR algorithms to compare later with the proposed CSSR algorithm.

- **Huber Markov Random Field (HMRF) Regulariser**

  It is a convex non-quadratic regulariser, which results in nonlinear cost functions, used to preserve edges, see [Schultz and Stevenson, 1996, Nasrollahi and Moeslund, 2014]. The energy function of an HMRF is defined as:

$$\rho(f_{\mathbf{x}}) = \begin{cases} f_{\mathbf{x}}^2, & if \mid f_{\mathbf{x}} \mid \leq \alpha \\ 2\alpha \mid f_{\mathbf{x}} \mid - \alpha^2, & otherwise \end{cases} \tag{3}$$

  where $f_{\mathbf{x}}$ is the first or second order differences of the HR image $\mathbf{x}$, α is a parameter separating the quadratic and the linear regions, that controls the size of discontinuities modelled by the regulariser by providing a less severe edge penalty. It has been used in [Schultz and Stevenson, 1996] to extract SR frames from video sequences.

- **Total Variation (TV) Regulariser**

The TV regulariser preserves image edges while smooths flat regions and avoids ringing artefacts [Chambolle et al., 2010]. The TV energy function is defined as

$$\rho(\mathbf{x}) = \| \nabla \mathbf{x} \|, \tag{4}$$

where ∇ is the gradient operator and $\| \cdot \|$ is the Euclidean norm. It can be approximated using

$$TV(\mathbf{x}) = \sum_{i=1}^{N} \sqrt{(\Delta_i^h(\mathbf{x}))^2 + (\Delta_i^v(\mathbf{x}))^2}\,, \tag{5}$$

where N is the number of pixels in $\mathbf{x}$, $\Delta_i^h(\mathbf{x})$ and $\Delta_i^h(\mathbf{x})$ correspond respectively to the horizontal and vertical first order differences at pixel i.

- **$\ell 1$ Norm Regulariser**

It is similar to the TV regulariser in its capability to preserve edges while imposing smoothness in the rest of image regions. Its energy function is defined as

$$\rho(\mathbf{x}) = \| \nabla \mathbf{x} \|_1, \tag{6}$$

where ∇ is the gradient operator and $\| \cdot \|_1$ is the $\ell 1$ norm, it can be expressed as

$$L1(\mathbf{x}) = \sum_{i=1}^{N} \alpha^h \| \Delta_i^h(\mathbf{x}) \|_1 + \sum_{i=1}^{N} \alpha^v \| \Delta_i^v(\mathbf{x}) \|_1\,, \tag{7}$$

where $\Delta_i^h(\mathbf{x})$ and $\Delta_i^v(\mathbf{x})$ represent the horizontal and vertical first order differences, respectively, at pixel i, and α^h, α^v are regulariser parameters. Notice here the use of two model parameters α^h and α^v to make this model more adaptable to image characteristics than the TV regulariser case. SR algorithm using $\ell 1$ prior, referred later as L1S, has been proposed in [Villena et al., 2009] which solves the minimisation problem in Equation (2).

- **The Simultaneous Autoregressive (SAR) Regulariser**

  It applies a uniform smoothness to all locations in the image. It is a non sparse prior that better reconstructs textures, see [Villena et al., 2013]. It uses the Euclidean norm of the second order difference, as follows

$$SAR(\mathbf{x}) = \| C\mathbf{x} \|^2 , \tag{8}$$

  where C is the Laplacian operator, and $\| \cdot \|$ is the Euclidean norm. An SR using SAR prior (referred later as SAS) algorithm, has been proposed in [Villena et al., 2013] which solves the minimisation problem in Equation (2).

- **Bilateral Total Variation (BTV) Regulariser**

  BTV preserves edges in the image, (see [Farsiu et al., 2004, Nelson et al., 2012, Nasrollahi and Moeslund, 2014]). It is expressed as:

$$BTV(\mathbf{x}) = \underbrace{\sum_{l=-P}^{P} \sum_{m=0}^{P}}_{l+m\geq 0} \alpha^{|m|+|l|} \| \mathbf{x} - \mathbf{S}_h^l \mathbf{S}_v^m \mathbf{x} \|_1 , \tag{9}$$

  where matrices $\mathbf{S}_h^l, \mathbf{S}_v^m$ shift the image $\mathbf{x}$ by l, m pixels in the horizontal and vertical directions, respectively; presenting several scales of derivatives. α is a positive parameter less than one, which gives more weight to closer pixels, $\| \cdot \|_1$ is the $\ell 1$ norm and P is a positive integer. The BTV regulariser has been used in [Farsiu et al., 2004]. Notice that when $P = 1$, the regulariser becomes the TV regulariser, and a comparison between the two regularisers was performed in [Farsiu et al., 2004].

  A fast and robust SR (FRSR) algorithm has been proposed in [Farsiu et al., 2004] which employs the $\ell 1$ norm minimisation in pursuit of robustness against motion errors and blurring. It uses the BTV regulariser. The following minimisation problem was solved to estimate $\mathbf{x}$

$$\hat{\mathbf{x}} = \arg\min_{\mathbf{x}} \left\{ \sum_q \| \mathbf{B}_q \mathbf{x} - \mathbf{y}_q \|_1 + \underbrace{\sum_{l=-P}^{P} \sum_{m=0}^{P}}_{l+m\geq 0} \alpha^{|m|+|l|} \| \mathbf{x} - \mathbf{S}_h^l \mathbf{S}_v^m \mathbf{x} \|_1 \right\} . \tag{10}$$

- **Generalised Gaussian MRF (GGMRF) Regulariser**

  It is a convex non-quadratic regulariser, see [Sauer et al., 1995, Nasrollahi and Moeslund, 2014]. It has the following energy function:

$$\rho(\mathbf{x}) = \mid \mathbf{x} \mid^{p}, \tag{11}$$

  where $1 < p < 2$. This regulariser tends to overly smooth the whole image including the edges, see [Capel and Zisserman, 2003], and it has been utilised in [Sauer et al., 1995] for tomographic image reconstruction problems.

The combination of some image regularisers in SR and image restoration has been addressed in [Villena et al., 2013, Starck et al., 2005, Villena et al., 2010]. In [Villena et al., 2013, Villena et al., 2010] sparse and non sparse priors were combined and applied to the SR image reconstruction problem.

A robust SR (RSR) Algorithm has been proposed in [Zomet et al., 2001] where instead of minimising the sum of the difference images, a median estimator is combined in an iterative process to obtain an SR algorithm. The HR image is estimated iteratively using the following update equation

$$\mathbf{x}^{k+1} = \mathbf{x}^{k} + Q \cdot \lambda \cdot median\{\mathbf{B}_q^t(\mathbf{B}_q\mathbf{x}^k - \mathbf{y}_q)\}_{q=1}^{Q}, \tag{12}$$

where k is the iteration number, Q is the number of LR input images, and λ is a scale factor.

In Section 4.1, we compare the CSSR algorithm with the L1S, SAS, FRSR, and RSR algorithms.

SR is different from, and expected to perform better than, interpolation techniques. Interpolation uses only one image to increase its spatial resolution, but it does not enhance the quality of the estimated image. On the contrary, it usually adds some smoothness and loss of some edges is expected. Examples of interpolation methods are: Bilinear interpolation where the output pixel value is a weighted average of the nearest four pixel values, and Bicubic interpolation where the neighbouring 16 pixels are used resulting in a smoother image with fewer interpolation artefacts, compared to the bilinear interpolation.

It is worth to discuss here the case where, in the model in Equation (1), both warping $\mathbf{C}(\mathbf{s}_q)$ and down-sampling matrices are the identity matrix, and so we only have the blur matrix $\mathbf{H}_q$, the modelling becomes

$$\mathbf{y}_q = \mathbf{H}_q\mathbf{x} + \mathbf{n}_q, \tag{13}$$

where again $\mathbf{y}_q$ is the observed image, $\mathbf{H}_q$ is the blur matrix applied to the original image $\mathbf{x}$, and $\mathbf{n}_q$ is the noise accompanying the blur operation. In this case, it is required to remove the blur from the observed image, and the problem becomes an Image Deconvolution (ID), or deblurring, problem. If $\mathbf{H}_q$ in the equation is assumed known then the problem falls into the Non-Blind Image Deconvolution (NBID), the more realistic case is where the matrix $\mathbf{H}_q$ has to be estimated, then the problem turns to be of the Blind Image Deconvolution (BID) type. Notice Equation (13) also describes the image deconvolution problem from multiple images [Šroubek and Flusser, 2003], if the deconvolution is made from only one single observation [Zhou et al., 2015], then the subscript q can be discarded.

ID is a post processing step to estimate a better quality image by restoring the high frequency content lost in the acquisition process. BID techniques can be included in the SR problem; by this the BID can form an additional step in SR post processing techniques. However, we will not go further in ID direction, as this work is devoted mainly to the SR problem from CS observations, assuming the blur is known.

SR has been already applied to many fields like high performance colour liquid crystal display screens, remote sensing, and medical imaging. Also, it has been used to enhance the resolution of still images from video sequences captured using conventional acquisition systems. For an extended list of applications using SR, refer to [Katsaggelos et al., 2007].

Finally, it is worth noting that the term SR has been used in other research areas as well. For example, in the field of optics, SR refers to a set of restoration procedures that seek to recover the information beyond the diffraction limit. In the scanning antenna research, SR techniques are exploited to resolve two closely spaced targets when a one-dimensional stepped scanning antenna is used [Tian and Ma, 2011]. It has also been used to refer to the problem of increasing the resolution of an image using learning techniques, a problem not addressed in this chapter.

2. Compressed Sensing

All classical compression techniques still obey the Nyquist's theorem, and must firstly sample the signal at a high rate and then compress it. Therefore the limitation on the sampling step still holds. Alternatively, CS is a new framework to sense in a compressed form data during the acquisition process itself, on sampling rates that can be lower than that of Nyquist's theorem. CS calculates projections of an image to represent it, rather than its original pixels.

CS, or compressed sampling, designs efficient sampling protocols to better capture only the useful information in an image utilising sparsity property. Assuming sparsity of an image in a transformed domain, then many transformed image components can be neglected as them introduce minimal information. As the significant components, those bearing most information, are not known a priori, a high incoherence between the sparse representation and the one utilised for sensing the image is required, [Candes and Wakin, 2008, Fornasier and Rauhut, 2011, Candes and Romberg, 2005].

Figure 3, shows the design of a typical CS image/video camera, designed by Rice University. The camera architecture employs a digital micromirror array to optically calculate linear projections of the scene onto pseudorandom binary patterns. Its key hallmark is its ability to obtain an image or video with a single detection element, a single pixel, while measuring the scene fewer times than the number of pixels/voxels.

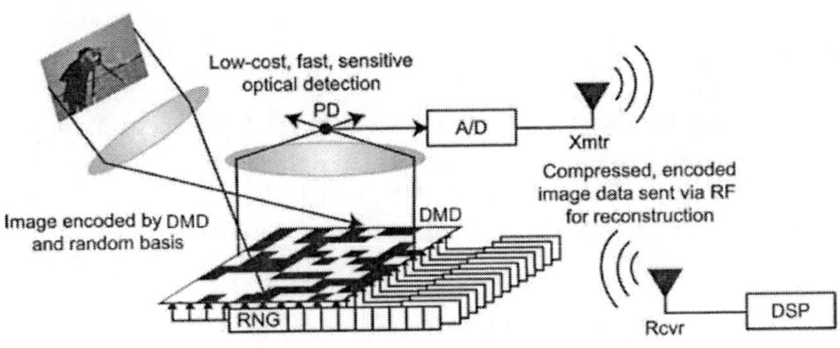

Figure 3. Typical CS camera.

To model the process, let us assume that we have an $P \times Q$ image. The rectangular matrix can be converted to the $N \times 1$ vector, $\mathbf{x}$, with $N = PQ$, as shown in Figure 4. The sampling, or measurement, process can be modelled by a 2-dimensional $M \times N$ matrix $\mathbf{\Phi}$, with $M \leq N$. The sensing process is equivalent to multiplying $\mathbf{\Phi}$ by $\mathbf{x}$ to give the $M \times 1$ observation vector $\mathbf{y}$. In the figure $M = 8$, $N = 16$, and the measurement matrix uses real valued entries, as will be explained later. CS theory establishes that a limited number of projections of the image, represented as entries of $\mathbf{y}$, are needed to recover the original image $\mathbf{x}$ efficiently. Mathematically,

$$\mathbf{y} = \mathbf{\Phi x} + \mathbf{n}, \tag{14}$$

where $\mathbf{n}$ represents the measurement noise assumed to be additive. Every entry in $\mathbf{y}$ is a projection of the original image, obtained by multiplying the corresponding row in $\mathbf{\Phi}$ by $\mathbf{x}$.

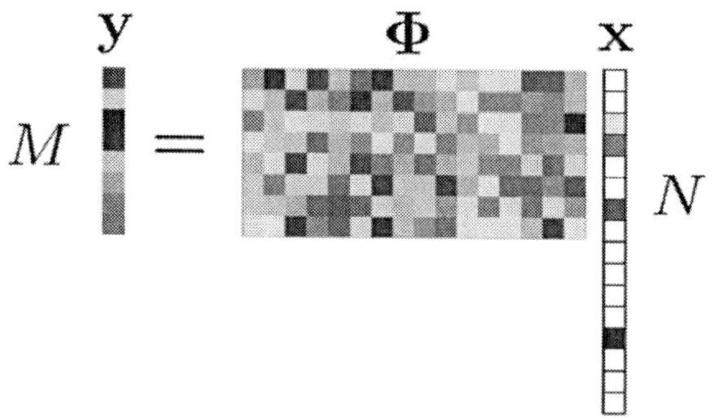

Figure 4. Compressed sensing acquisition model. Measurement matrix $\mathbf{\Phi}$ with real entries, $M = 8$, $N = 16$.

To recover the image $\mathbf{x}$ from the observation $\mathbf{y}$, CS theory relies on the idea that the recovered image should be sparse in a transformed domain, that is, we can use the transform basis $\mathbf{W}$ to find the transform coefficients $\mathbf{a}$ such that

$$\mathbf{x} = \mathbf{Wa}\,, \tag{15}$$

where $\mathbf{W}$ is an $N \times N$ matrix, and $\mathbf{x}$ is said to be S-sparse if only S of the $\mathbf{a}$ components of its representation in the $\mathbf{W}$ basis are non vanishing. Then Equation (14) becomes

$$\mathbf{y} = \mathbf{\Phi Wa} + \mathbf{n}\,. \tag{16}$$

The coherence μ, of a CS system measures the maximum correlation between the measurement basis and transform basis vectors. It can be expressed as:

$$\mu = \sqrt{N} \max_{k,j} \mid \mathbf{\Phi}_k \mathbf{W}_j \mid, \tag{17}$$

where $\mathbf{\Phi}_k$ represents the $k-$th vector of the $\mathbf{\Phi}$ basis, $\mathbf{W}_j$ represents the $j-$th column in $\mathbf{W}$ with $1 \leq j \leq N$. In Equation (17), if matrices used were in normalised form, which is usually the case, then the range of possible values for μ is $1 \leq \mu \leq \sqrt{N}$. For a CS theory efficient application, CS systems coherence coefficient values μ have to be low.

A measurement matrix $\mathbf{\Phi}$ obeys the Restricted Isometry Property (RIP) of order S if the isometry constant δ_S of $\mathbf{\Phi}$ is not too close to one for each integer $S = 1, 2, \cdots$. The δ_S is defined as the smallest number such that the following holds for all S-sparse vectors of $\mathbf{x}$

$$(1 - \delta_S) \parallel \mathbf{x} \parallel^2 \leq \parallel \mathbf{\Phi x} \parallel^2 \leq (1 + \delta_S) \parallel \mathbf{x} \parallel^2 \tag{18}$$

This is equivalent also to saying that all subsets of S columns taken from $\mathbf{\Phi}$ are nearly orthogonal. Notice that the columns cannot be exactly orthogonal since number of columns usually is larger than number of rows in $\mathbf{\Phi}$, see [Candes and Wakin, 2008, Candes and Tao, 2005].

A random sensing matrix $\mathbf{\Phi}$ and a basis $\mathbf{W}$ allowing an S-sparse representation of $\mathbf{x}$ will have the RIP if the following expression is accomplished

$$M \geq c \cdot S \cdot \log(N/S), \tag{19}$$

where c is a constant. This ensures that the reconstruction of an $\mathbf{x}$ image with an $S-$sparse representation in a vector basis $\mathbf{W}$ can be obtained from a compressed sensing observation $\mathbf{y}$ through the following minimisation problem.

$$\hat{\mathbf{a}} = \arg\min_{\mathbf{a}} \parallel \mathbf{a} \parallel_0 \text{ subject to } \mathbf{a} : \parallel \mathbf{\Phi W a} - \mathbf{y} \parallel^2 \leq \epsilon, \tag{20}$$

where the l_0 norm $\parallel \mathbf{a} \parallel_0$ counts the number of nonzero entries in a vector $\mathbf{a}$, $\parallel \cdot \parallel$ is the Euclidean norm, and ϵ is a small value that determines the accuracy of the estimation process. While the Euclidean norm is convex, the zero norm is non convex, that makes the problem very difficult to solve. A way to make the problem more tractable is to replace the zero norm with its convex approximation $\parallel . \parallel_1$, that is the $\ell 1$ norm. Hence the problem becomes,

$$\hat{\mathbf{a}} = \arg\min_{\mathbf{a}} \parallel \mathbf{a} \parallel_1 \text{ subject to } \mathbf{a} : \parallel \mathbf{\Phi W a} - \mathbf{y} \parallel^2 \leq \epsilon, \tag{21}$$

All constituting terms are convex now, hence the minimisation problem is computationally feasible. Moreover, there are a variety of reasons which justify the use of $\ell 1$-minimisation for sparse signal recovery, see [Davenport et al., 2011]. Converting the constrained optimisation to a none constrained one gives,

$$\hat{\mathbf{a}} = \arg\min_{\mathbf{a}} \frac{\eta}{2} \parallel \mathbf{\Phi W a} - \mathbf{y} \parallel^2 + \tau \parallel \mathbf{a} \parallel_1, \tag{22}$$

where η, τ are positive parameters. This is the minimisation problem used in the CSSR which has been also utilised in [Kim et al., 2007, Candes and Romberg, 2005, Amizic et al., 2013b].

Suitable transform bases ($\mathbf{W}$) that serve well in the CS reconstruction are Discrete Cosine Transform (DCT), Discrete Wavelet Transform (DWT) and Discrete Fourier Transform (DFT). DWT is usually preferred over DCT because it better enables the removal of blocking artefacts [Jeengar et al., 2012]. In this work we adapted interchangeably the DCT, and the DWT. Both are described next.

DCT attempts to make use of correlated data in the image $\mathbf{z}_q$. In the one-dimensional (1D) DCT the signal is converted into a sum of cosine functions oscillating at different frequencies. It can be expressed as follows

$$\mathbf{a}_q(k) = \boldsymbol{\omega}(k) \sum_{i=1}^{D} \mathbf{z}_q(i) \cos\left[\frac{\pi}{2D}(2i-1)(k-1)\right], \quad k = 1, \cdots, D, \tag{23}$$

where $\mathbf{z}_q$ is expressed in the vector form, and

$$\boldsymbol{\omega}(k) = \begin{cases} \frac{1}{\sqrt{D}}, & k = 1 \\ \sqrt{\frac{2}{D}}, & 2 \leq k \leq D \end{cases} \tag{24}$$

where D is the number of pixels in $\mathbf{z}_q$. The inverse 1D-DCT reconstructs a sequence from its transform coefficients by

$$\mathbf{z}_q(i) = \sum_{k=1}^{D} \boldsymbol{\omega}(k) \mathbf{a}_q(k) \cos\left[\frac{\pi}{2D}(2i-1)(k-1)\right], \quad i = 1, \cdots, D, \tag{25}$$

DWT invokes two types of filters; the wavelet filter and the scaling vector. The wavelet filter is a high pass filter, and the scaling vector is a low pass filter. After applying a 1-level Haar DWT on an image $\mathbf{z}_q$ in the matrix form, we decompose it into four frequency bands, namely the approximation (LL_1), horizontal (LH_1), vertical (HL_1), and diagonal (HH_1) sub-bands. This DWT decomposition at the 1-level is shown in Figure 5 for Barbara image to represent the image $\mathbf{z}_q$. The original image is shown in Figure 5(a), while Figure 5(b) shows the four down-sampled sub-bands: approximation, horizontal, vertical and diagonal bands. Notice that if we append the four bands together in one vector we obtain the 1-level coefficient vector which has the same size as $\mathbf{z}_q$.

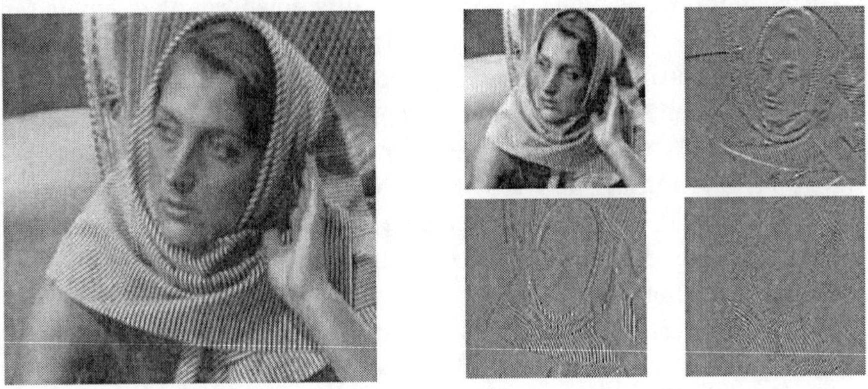

(a) (b)

Figure 5. 1-level discrete wavelet transform. (a) Barbara image (b) DWT decomposition.

If the 1-level DWT is applied again to the approximation sub-band LL_1, then we obtain the 2-level DWT, and hence we will have the new sub-bands LL_2, LH_2, HL_2, and HH_2. Applying DWT again to the resulting approximation sub-band LL_2, then we will have the 3-level DWT, with the new added bands LL_3, LH_3, HL_3 and HH_3. This work used 3-level Haar DWT as the wavelet transform domain.

CS is a promising technology which decreases the acquisition time and memory size by using only one sensor. The use of a single sensor makes it suitable for imaging at wavelengths where CCD and CMOS are blind [Candes and Wakin, 2008].

Different CS image/video cameras have been proposed [Duarte et al., 2008, Chan et al., 2008, Heidari and Saeedkia, 2009, Li et al., 2013, Sankaranarayanan et al., 2012]. The concept of single pixel camera has been extended to a multi-detector device in [Edeler et al., 2011], where the detectors are placed on an arbitrary grid on the source signal and each detector integrates the source signals only locally to reduce the required measurements, and hence the measurement time, by using more detectors.

SR from a single image has benefited from the introduction of CS theory in works like [Fannjiang and Liao, 2012, Babcock et al., 2013, Fang et al., 2014]. In [He et al., 2012, Yang et al., 2011] learning based SR is used to estimate an HR image from a single CS observation of a down-sampled remote sensing image. In [Sen and Darabi, 2009] the down-sampling is incorporated in the measurement matrix, the CS image is reconstructed in the wavelet domain and the signal is deconvolved in the Fourier domain.

BID from CS measurement has been studied in [Spinoulas et al., 2012, Amizic et al., 2013a], and NBID in [Koh et al., 2007, Figueiredo et al., 2007, Needell and Tropp, 2009, Becker et al., 2011, Yang and Zhang, 2011]. Next we describe selected CS ID works which will be compared with the CSSR algorithm in Section 4.2.

The Large Scale $\ell 1$ regularised (L1-LS) logistic regression algorithm proposed in [Koh et al., 2007], and Gradient Projection for Sparse Reconstruction (GPSR) algorithm proposed in [Figueiredo et al., 2007], both solve the following problem

$$\hat{\mathbf{a}} = \arg\min_{\mathbf{a}} \parallel \mathbf{\Phi}\mathbf{H}\mathbf{W}\mathbf{a} - \mathbf{y} \parallel^2 + \tau \parallel \mathbf{a} \parallel_1 , \tag{26}$$

where $\mathbf{y}$ is the CS blurred observation, $\mathbf{\Phi}$ the measurement matrix, $\mathbf{H}$ the blur matrix, $\mathbf{W}$ is the transform domain basis, $\mathbf{a}$ is the transformed domain representation of the original image $\mathbf{x} = \mathbf{W}\mathbf{a}$ and τ is a non-negative parameter.

The Compressive Sampling Matching Pursuit (CoSaMP) algorithm proposed in [Needell and Tropp, 2009] is based on Orthogonal Matching Pursuit (OMP) schemes and produces an $s-$sparse approximation $\mathbf{z}$ that satisfies

$$\parallel \mathbf{x} - \mathbf{z} \parallel_2 \leq C \cdot \max\left\{\eta , \frac{1}{\sqrt{s}} \parallel \mathbf{x} - \mathbf{x}_{s/2} \parallel_1 + \parallel \mathbf{y} - \mathbf{\Phi}\mathbf{z} \parallel_2 \right\} , \tag{27}$$

where C and η are parameters, and $\mathbf{x}_{s/2}$ is the best $(s/2)-$sparse approximation to $\mathbf{x}$.

The NESTerov's Algorithm (NESTA) proposed in [Becker et al., 2011] and Your ALgorithm for L1 (YALL1) proposed in [Yang and Zhang, 2011] solve the following problem

$$\hat{\mathbf{x}} = \arg\min_{\mathbf{x}} \parallel \mathbf{W}^{-1}\mathbf{x} \parallel_1 \quad \text{s.t.} \quad \parallel \mathbf{\Phi}\mathbf{H}\mathbf{x} - \mathbf{y} \parallel < \epsilon , \tag{28}$$

3. CSSR Modelling and Formulation

In the Compressed Sensing Super Resolution (CSSR) problem, we are given multiple CS LR observations of a scene to estimate an HR image. The LR images can be considered as a warped, blurred, and down-sampled versions of the HR image to be estimated. Let us assume that the number of input images is Q, the acquisition model for each observation is

$$\mathbf{y}_q = \mathbf{\Phi}\mathbf{z}_q + \mathbf{r}_q \quad q = 1, \ldots, Q, \tag{29}$$

where $\mathbf{y}_q$ is an $M \times 1$ vector representing the compressed observations, $\mathbf{\Phi}$ is a CS $M \times D$ measurement matrix, $\mathbf{z}_q$ is a column vector of size $D \times 1$ representing the $q-$th LR image and $\mathbf{r}_q$ represents the observation noise. We denote by R the compression ratio of the measurement system, that is, $R = M/D$, $R \leq 1$.

As shown in Section 1 the observation process of the LR observations $\mathbf{z}_q$ in Equation (29) can be modelled as follows,

$$\mathbf{z}_q = \mathbf{A}\mathbf{H}_q\mathbf{C}(\mathbf{s}_q)\mathbf{x} + \mathbf{w}_q = \mathbf{B}_q(\mathbf{s}_q)\mathbf{x} + \mathbf{w}_q, \tag{30}$$

where $\mathbf{A}$ is a $D \times N$ down-sampling matrix, $D \leq N$, which models the limited resolution of the acquisition system, when capturing the high resolution image, where $N = P^2 D$ and P$\geq$1 is the zooming factor, in each dimension of the image. $\mathbf{H}_q$ is an $N \times N$ blurring matrix, modelling the blur action accompanying the imaging process. In this work, the convolution matrix $\mathbf{H}_q$ is assumed to be known. $\mathbf{C}(\mathbf{s}_q)$ is the $N \times N$ warping matrix for the motion vector $\mathbf{s}_q = [\theta_q, c_q, d_q]^t$, where θ_q is the rotation angle, c_q and d_q are respectively the horizontal and vertical translations of the $q-$th LR image with respect to the reference frame, and $\mathbf{x}$ is the HR image of size $N \times 1$. Finally, $\mathbf{w}_q$ models the noise associated to the LR acquisition process. We write $\mathbf{B}_q(\mathbf{s}_q) = \mathbf{A}\mathbf{H}_q\mathbf{C}(\mathbf{s}_q)$ for simplicity.

Using Equations (29) and (30) we can write

$$\mathbf{y}_q = \mathbf{\Phi}\mathbf{B}_q(\mathbf{s}_q)\mathbf{x} + \mathbf{n}_q\,, \text{ for } q = 1, \ldots, Q, \tag{31}$$

where $\mathbf{n}_q$ represents the combined CS and LR acquisition noise and $\mathbf{x}$ is the HR image we want to estimate.

Next we describe the matrices used in this model. The sensing matrix $\mathbf{\Phi}$ used in this work is a random circulant Toeplitz matrix. A circulant Toeplitz matrix is a matrix $\mathbf{T}_n$, in which each row is a cyclic shift of the row above. It has the following form

$$\mathbf{T}_n = \begin{bmatrix} t_0 & t_1 & t_2 & \cdots & t_{n-1} \\ t_{n-1} & t_0 & t_1 & & t_{n-2} \\ t_{n-2} & t_{n-1} & t_0 & & \\ \vdots & & & \ddots & \\ t_1 & & & & t_0 \end{bmatrix} \tag{32}$$

The entries in the first row are random, following a Gaussian, or Bernoulli distribution. Gaussian, or normal, distribution function is given by

$$f(x|\mu,\sigma) = \frac{1}{\sigma\sqrt{2\pi}} e^{-\frac{(x-\mu)^2}{2\sigma^2}} \tag{33}$$

where μ and σ^2 are respectively the mean and the standard deviation of the distribution. Bernoulli probability function is given by

$$f(k|p) = p^k(1-p)^{1-k}, \quad \text{for } k \in \{0,1\}, \tag{34}$$

where (p) and $(1-p)$ are respectively the probabilities of k being 1 and 0.

Both Gaussian and Bernoulli random circulant Toeplitz matrices were tested to work well in CS [Bajwa et al., 2007, Zhang et al., 2010, Yin et al., 2010]. In both cases the rows of $\mathbf{\Phi}$ are normalised to 1. However, a Gaussian matrix consists of real entries while a Bernoulli matrix consists of binary entries, and hence it can be synthesised practically, see [Heidari and Saeedkia, 2009, AlSaafin et al., 2016]. This work used the Bernoulli circulant Toeplitz matrix.

Next we describe the down-sampling matrix $\mathbf{A}$. Let us assume that we have an image $\mathbf{x}$ of size 256×256, and the zooming factor P=2. We divide the matrix $\mathbf{x}$ into $P \times P$ blocks, then from each block only one pixel (the top-left in the block) value is preserved, and the other three are discarded. The preserved pixels constitute a 128×128 matrix, which is the down-sampled version of $\mathbf{x}$. Notice here that there is no interpolation process neither in the down-sampling nor in the up-sampling process. If the down-sampled 128×128 matrix in the previous example is up-sampled, then each pixel value will form a block in the up sampled image, the upper-left pixel of the four pixels forming this block will have the pixel value from the down-sampled image, the other three pixels will have a zero value.

The blur matrix $\mathbf{H}$ we used in all experiments utilises a spatially invariant Gaussian blur filter given by

$$h(\mathbf{u}) = K\, e^{-\frac{|\mathbf{u}|^2}{2\sigma^2}} \tag{35}$$

where K is a constant and σ^2 is the blur variance.

To calculate the warping matrix $\mathbf{C}(\mathbf{s}_q)$ let us denote the coordinates of the reference HR grid by (u, v) and the coordinates of the $q - th$ warped HR grid, after applying $\mathbf{C}(\mathbf{s}_q)$ to $\mathbf{x}$, by (u_q, v_q). Then it holds that

$$u_q = u\cos(\theta_q) - v\sin(\theta_q) + c_q \tag{36}$$

$$v_q = u\sin(\theta_q) + v\cos(\theta_q) + d_q\,. \tag{37}$$

Let us denote the displacements between the grids by $\Delta(u_q,v_q)^T = (u, v)^T - (u_q, v_q)^T$. The vector difference between the pixel at (u_q, v_q) and the pixel at its top-left position in the reference HR grid is denoted by $(a_q(\mathbf{s}_q), b_q(\mathbf{s}_q))^T$ (see Figure 6), that is,

$$a_q(\mathbf{s}_q) = \Delta u_q - \mathrm{floor}(\Delta u_q)\,, \tag{38}$$

$$b_q(\mathbf{s}_q) = \Delta v_q - \mathrm{floor}(\Delta v_q)\,. \tag{39}$$

Notice here that the relationships are non linear functions of $\mathbf{s}_q$ due to the trigonometric functions in Equations (36) and (37). Using bilinear interpolation, the warped image $\mathbf{C}(\mathbf{s}_q)\mathbf{x}$ can be approximated to obtain a linear transform matrix in $\mathbf{s}_q$ as (see [He et al., 2007, AlSaafin et al., 2016, Villena et al., 2013]),

$$\begin{aligned}\mathbf{C}(\mathbf{s}_q)\mathbf{x} \approx\, & \mathbf{D}_{b_q(\mathbf{s}_q)}(\mathbf{I} - \mathbf{D}_{a_q(\mathbf{s}_q)})\mathbf{L}_{\mathbf{bl}(\mathbf{s}_q)}\mathbf{x} + (\mathbf{I} - \mathbf{D}_{b_q(\mathbf{s}_q)})\mathbf{D}_{a_q(\mathbf{s}_q)}\mathbf{L}_{\mathbf{tr}(\mathbf{s}_q)}\mathbf{x} \\ & + (\mathbf{I} - \mathbf{D}_{b_q(\mathbf{s}_q)})(\mathbf{I} - \mathbf{D}_{a_q(\mathbf{s}_q)})\mathbf{L}_{\mathbf{tl}(\mathbf{s}_q)}\mathbf{x} + \mathbf{D}_{b_q(\mathbf{s}_q)}\mathbf{D}_{a_q(\mathbf{s}_q)}\mathbf{L}_{\mathbf{br}(\mathbf{s}_q)}\mathbf{x}\,,\end{aligned} \tag{40}$$

where $\mathbf{D}_{a_q(\mathbf{s}_q)}$ and $\mathbf{D}_{b_q(\mathbf{s}_q)}$ denote diagonal matrices with the vectors $a_q(\mathbf{s}_q)$ and $b_q(\mathbf{s}_q)$ in their diagonals, respectively. $\mathbf{I}$ is the identity matrix. Matrices $L_{\mathbf{z}}$ with $\mathbf{z} \in \{\mathbf{bl}(\mathbf{s}_q), \mathbf{br}(\mathbf{s}_q), \mathbf{tl}(\mathbf{s}_q), \mathbf{tr}(\mathbf{s}_q)\}$ are constructed in such a way that the product $L_{\mathbf{z}}\mathbf{x}$ produces pixels at the bottom-left, bottom-right, top-left, and top-right, locations of (u_q, v_q), respectively. This approximated explicit form presented in Equation (40) can be used when we are given the motion vectors of the input LR CS observations.

The noise matrix $\mathbf{n}_q$ used in all experiments is assumed to be white Gaussian noise of zero mean, added to the CS observations. This noise arises during

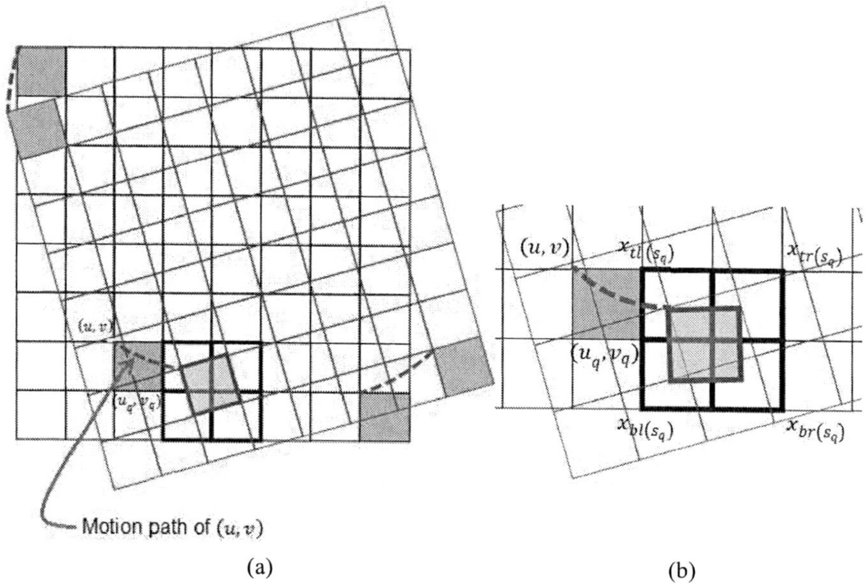

Figure 6. q-th HR grid calculation. (a) HR grid (in black) and the q-th image grid (in red), (b) Detailed view of (a), with the pixel notation used for the bilinear interpolation of grid element (u_q, v_q).

acquisition due to poor illumination, high temperature and electronic circuit noise, see [Farooque and Rohankar, 2013].

Let us now show an example to clarify visually the presented model. Figure 7(a) shows the Shepp-Logan image. The original image $\mathbf{x}$ is warped then blurred then down-sampled, following Equation (30) to get $\mathbf{B}_q(\mathbf{s}_q)\mathbf{x}$ for the case $Q = 4$. In Figure 7(b) the warping used the following motion vectors

$$\mathbf{s}_1 = \begin{bmatrix} 0 \\ 0 \\ 0 \end{bmatrix}, \mathbf{s}_2 = \begin{bmatrix} 0.1571 \\ 2 \\ -3 \end{bmatrix}, \mathbf{s}_3 = \begin{bmatrix} -0.1920 \\ -1 \\ -2 \end{bmatrix}, \text{ and } \mathbf{s}_4 = \begin{bmatrix} 0.0524 \\ 13 \\ -7 \end{bmatrix}. \quad (41)$$

To blur the warped image we used a Gaussian blur with variance 5, the down-sampling used a zooming factor P=2. The four generated LR images $(\mathbf{B}_q(\mathbf{s}_q)\mathbf{x})$ are shown in Figure 7(b). The simulated LR images then were compressed to get $\mathbf{y}_q$ as in Equation (31). Firstly the image is compressed using a compression ratio R=0.6 then noised by an additive white noise of SNR=30dB. The four noised CS observations are illustrated into a 2D form in Figure 7(c).

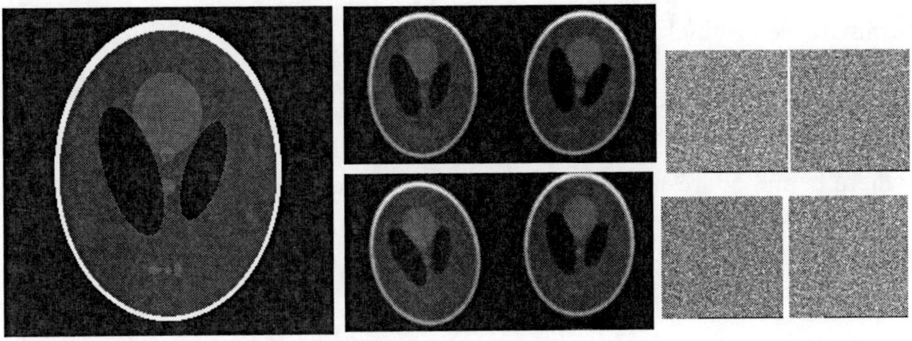

(a) (b) (c)

Figure 7. Simulation process. (a) Original Shepp-Logan, (b) Four LR images (Warping with motion vectors $[0, 0, 0]^t$, $[.1571, 2, -3]^t$, $[-.1920, -1, -2]^t$, $[.0524, 13, -7]^t$, Gaussian blur of variance 5, zooming factor 2), (c) Four CS observations (compression ratio R=0.6, SNR of added noise 30dB).

The CSSR aims at finding an estimate for the HR image in Figure 7(a) given some noised CS observations, like those in Figure 7(c). Next we formulate the CSSR minimisation problem.

Since $\mathbf{z}_q$ in Equation (29) represents translated and rotated LR versions of the original image $\mathbf{x}$, which are assumed to be compressible in a transformed domain, we can estimate the original HR image by first recovering the LR images using CS techniques and then recover the HR image using standard SR techniques on the recovered LR images. To be precise, if we assume that the LR images are sparse in a transformed domain with $\mathbf{z}_q$=$\mathbf{W}\mathbf{a}_q$, where $\mathbf{W}$ is a sparse promoting transformation of size $D \times D$ and $\mathbf{a}_q$ is the $q-$th transformed coefficient vector, we want to recover all $\mathbf{z}_q$ using a CS reconstruction technique.

The CS recovery methods available usually utilise the $\ell 1$ minimisation or Greedy methods. Examples on Greedy methods are Matching Pursuit (MP), Orthogonal MP (OMP), Subspace Pursuit (SP), regularised OMP (ROMP), Iterative Hard Shareholding (IHT) algorithm, see [Meenakshi, 2015, Huang and Makur, 2011, Tropp and Gilbert, 2007, Dai and Milenkovic, 2009, Needell and Vershynin, 2010] for details. The CSSR uses $\ell 1$ minimisation in Equation (22) to recover the $\mathbf{z}_q$ in Equation (29). Then defining $\hat{\mathbf{z}}_q = \mathbf{W}\hat{\mathbf{a}}_q$ and $\mathbf{s} = (\mathbf{s}_1, \ldots, \mathbf{s}_Q)$ and using the degradation model in Equation (30), we can estimate the original image by solving

$$\hat{\mathbf{x}}, \hat{\mathbf{s}} = \arg\min_{\mathbf{x},\mathbf{s}} \frac{\beta}{2} \sum_q \| \mathbf{B}_q(\mathbf{s}_q)\mathbf{x} - \hat{\mathbf{z}}_q \|^2 + \alpha\, \mathbf{Q}(\mathbf{x}), \tag{42}$$

where α and β are non-negative parameters, and the term $\mathbf{Q}(\mathbf{x})$ is described next. In [AlSaafin et al., 2016, Saafin et al., 2017] the following regularisation term was used for the image

$$\mathbf{Q}(\mathbf{x}) = \sum_{d\in\Delta} \sum_{i=1}^{N} \log(|\omega_d^{\mathbf{x}}(i)|)\,, \tag{43}$$

where $\omega_d^{\mathbf{x}}(i)$ is the i-th pixel of the filtered image, and

$$\omega_d^{\mathbf{x}} = \mathbf{F}_d\mathbf{x}\,, \tag{44}$$

where $\mathbf{F}_d$ is a high-pass filter operator, and the index $d \in \Delta$ identifies one of the members of the used filter set. This work used a filter set with elements $\Delta = \{h, v, hv, vh, hh, vv\}$, where h, v represent the first order horizontal and vertical difference filters, hv and vh represent first order differences along diagonals, and hh and vv the horizontal and vertical second order differences. The regularisation term favors sparsity of the high-pass filtered images $\mathbf{F}_d\mathbf{x}$, and corresponds to the Super-Gaussian log prior used in blind deconvolution [Babacan et al., 2012].

Since the log function can not be differentiated at zero, the CSSR considers the following robust version of the log regulariser

$$\log_\epsilon(|s|) = \begin{cases} \log(|s|) & \text{for } |s| \geq \epsilon \\ \frac{s^2}{2\epsilon^2} - (\frac{1}{2} - \log(\epsilon)) & \text{for } 0 \leq |s| \leq \epsilon \end{cases} \tag{45}$$

and replace $\mathbf{Q}(\mathbf{x})$ in Equation (43) by

$$\mathbf{Q}(\mathbf{x}) = \sum_{d\in\Delta} \sum_{i=1}^{N} \log_\epsilon(|\omega_d^{\mathbf{x}}(i)|)\,. \tag{46}$$

We have two approaches to solve the CSSR problem: the sequential approach and the alternate approach. The latter approach alternates between compressive sensing reconstruction, registration parameter estimation, and reconstruction of the HR image. The former approach estimates the unknowns sequentially, one after the other, as follows. Firstly, the LR images are reconstructed using Equation (22), then motion parameters are estimated and finally the HR image is estimated using Equation (42). Notice here that after an unknown is estimated, the sequential approach does not return back to it again, and every unknown is estimated only once.

As we will show in the experimental part, combining the two optimisation problems in Equations (22) and (42) above into a simultaneous one leads to an improved performance, as this enables better exploitation of the compressibility of the LR observations using the additional information obtained from the estimated HR image, and the updated registration parameters. Hence, in the following, this alternate approach has been adopted. According to it, let $\mathbf{a} = (\mathbf{a}_1, \ldots, \mathbf{a}_Q)$ and define

$$L(\mathbf{x}, \mathbf{a}) = \frac{\eta}{2} \sum_{q=1}^{Q} \| \mathbf{\Phi} \mathbf{W} \mathbf{a}_q - \mathbf{y}_q \|^2 + \tau \sum_{q=1}^{Q} \| \mathbf{a}_q \|_1 + \alpha \, \mathbf{Q}(\mathbf{x}) \tag{47}$$

Then we approach the CSSR problem by solving the following constrained optimisation problem

$$\begin{aligned} &\min L(\mathbf{x}, \mathbf{a}) \\ &\text{s.t.} \quad \mathbf{B}_q(\mathbf{s}_q)\mathbf{x} = \mathbf{W}\mathbf{a}_q \, , \text{for } \; q = 1, \ldots, Q \, . \end{aligned} \tag{48}$$

This constrained optimisation can be changed to be unconstrained using Alternate Direction Method of Multipliers (ADMM) algorithm. The new augmented Lagrangian functional is defined as follows

$$\begin{aligned} L(\mathbf{x}, \mathbf{a}, \mathbf{s}, \boldsymbol{\lambda}) = {} & \frac{\eta}{2} \sum_{q=1}^{Q} \| \mathbf{\Phi} \mathbf{W} \mathbf{a}_q - \mathbf{y}_q \|^2 + \tau \sum_{q=1}^{Q} \| \mathbf{a}_q \|_1 + \alpha \, \mathbf{Q}(\mathbf{x}) \\ & + \sum_{q=1}^{Q} \boldsymbol{\lambda}_q^t (\mathbf{B}_q(\mathbf{s}_q)\mathbf{x} - \mathbf{W}\mathbf{a}_q) + \frac{\beta}{2} \sum_{q=1}^{Q} \| \mathbf{B}_q(\mathbf{s}_q)\mathbf{x} - \mathbf{W}\mathbf{a}_q \|^2, \end{aligned} \tag{49}$$

where $\mathbf{s} = (\mathbf{s}_1, \ldots, \mathbf{s}_Q)$ and $\boldsymbol{\lambda} = (\boldsymbol{\lambda}_1, \ldots, \boldsymbol{\lambda}_Q)$ denote the motion vector and Lagrangian multiplier sets, respectively. The HR regularisation term $\mathbf{Q}(\mathbf{x})$ utilised in the CSSR algorithm is described in Equation (46).

The augmented Lagrangian functional in Equation (49) and the ADMM lead to the following iterative unconstrained sub-problems;

$$\mathbf{x}^{k+1} = \arg\min_{\mathbf{x}} L(\mathbf{x}, \mathbf{a}^k, \mathbf{s}^k, \boldsymbol{\lambda}^k), \tag{50}$$

$$\mathbf{a}^{k+1} = \arg\min_{\mathbf{a}} L(\mathbf{x}^{k+1}, \mathbf{a}, \mathbf{s}^k, \boldsymbol{\lambda}^k) \tag{51}$$

$$\mathbf{s}^{k+1} = \arg\min_{\mathbf{s}} L(\mathbf{x}^{k+1}, \mathbf{a}^{k+1}, \mathbf{s}, \boldsymbol{\lambda}^k) \tag{52}$$

$$\boldsymbol{\lambda}_q^{k+1} = \boldsymbol{\lambda}_q^k - \beta[\mathbf{B}_q(\mathbf{s}_q^{k+1})\mathbf{x}^{k+1} - \mathbf{W}\mathbf{a}_q^{k+1}], \quad q = 1, \dots, Q, \tag{53}$$

where k is the iteration number. Notice that according to the ADMM approach $\mathbf{B}_q(\mathbf{s}_q)$ in Equation (49) should not depend on the iteration index, as it is not the case here. However, we have not encountered any convergence problems applying this iterative procedure.

The calculation of the $\boldsymbol{\lambda}_q^{k+1}$ is straightforward. The estimation process of the other unknowns is described in the coming subsections.

3.1. HR Image Estimation

The function $\rho_\epsilon(s) = \log_\epsilon(|s|)$ in Equation (46) is symmetric around 0, and $\rho(\sqrt{s})$ is concave and increasing for $s \in [0, \infty)$ [Babacan et al., 2012]. So, it can be represented as (see [Rockafellar, 1997])

$$\rho_\epsilon(s) = \inf_{\xi>0} \frac{1}{2}\xi s^2 - \rho_\epsilon^*(\frac{1}{2}\xi), \tag{54}$$

where $\rho_\epsilon^*(\frac{1}{2}\xi)$ is the concave conjugate function

$$\rho_\epsilon^*(\frac{1}{2}\xi) = \inf_{s>0} \frac{1}{2}\xi s^2 - \rho_\epsilon(s). \tag{55}$$

It is shown in [Babacan et al., 2012] that the infimum in Equation (54) is achieved when $\xi = \rho'_\epsilon(s)/s$. Consequently, for the regularisation term $\mathbf{Q}(\mathbf{x})$ in Equation (49), we can write

$$\mathbf{Q}(\mathbf{x}) \le R(\mathbf{x}, \boldsymbol{\xi}) = \frac{1}{2}\sum_{d\in\Delta} \mathbf{x}^t\mathbf{F}_d^t\boldsymbol{\Omega}_d\mathbf{F}_d\mathbf{x} - \sum_{d\in\Delta}\sum_{i=1}^{N} \rho_\epsilon^*(\frac{1}{2}\xi_d(i)) \tag{56}$$

where $\boldsymbol{\xi} = (\boldsymbol{\xi}_1, \dots, \boldsymbol{\xi}_Q)$, $\boldsymbol{\xi}_q = (\xi_q(1), \dots, \xi_q(N))$ for $q = 1, \dots, Q$, with all its components positive, and $\boldsymbol{\Omega}_d$ is a diagonal matrix with entries

$$\Omega_d(i, i) = \xi_d(i). \tag{57}$$

For a given $\mathbf{x}$, the inequality in Equation (56) becomes an equality if

$$\xi_d^{\mathbf{x}}(i) = \min(1/|\omega_d^{\mathbf{x}}(i)|^2, 1/\epsilon^2) = \begin{cases} \frac{1}{|\omega_d^{\mathbf{x}}(i)|^2} & \text{for } |\omega_d^{\mathbf{x}}(i)| \geq \epsilon \\ \frac{1}{\epsilon^2} & \text{for } 0 \leq |\omega_d^{\mathbf{x}}(i)| \leq \epsilon \end{cases} \tag{58}$$

where $\omega_d^{\mathbf{x}}(i)$ is defined from $\mathbf{x}$ in Equation (44). Then we can apply standard Majorisation-Minimisation methods [Lange, 2013]. Given $\mathbf{x}^k$, $\mathbf{a}^k$, $\mathbf{s}^k$ and defining

$$L^k(\mathbf{x}) = \frac{\beta}{2}\sum_q \| \mathbf{B}_q(\mathbf{s}_q^k)\mathbf{x} - \mathbf{W}\mathbf{a}_q^k \|^2 + \sum_q \boldsymbol{\lambda}_q^{k^t}(\mathbf{B}_q(\mathbf{s}_q^k)\mathbf{x} - \mathbf{W}\mathbf{a}_q^k) \tag{59}$$

then

$$L^k(\mathbf{x}^k) + \alpha\mathbf{Q}(\mathbf{x}^k) \geq L^k(\mathbf{x}^{k+1}) + \alpha\mathbf{Q}(\mathbf{x}^{k+1}) \tag{60}$$

where

$$\mathbf{x}^{k+1} = \arg\min_{\mathbf{x}} \left\{ \frac{\beta}{2}\sum_q \| \mathbf{B}_q(\mathbf{s}_q^k)\mathbf{x} - \mathbf{W}\mathbf{a}_q \|^2 + \alpha R(\mathbf{x}, \boldsymbol{\xi}^{\mathbf{x}^k}) + \sum_q \boldsymbol{\lambda}_q^{k^t}(\mathbf{B}_q(\mathbf{s}_q^k)\mathbf{x} - \mathbf{W}\mathbf{a}_q^k) \right\}. \tag{61}$$

From Equation (61), the optimisation step in Equation (50) produces the following linear equation for $\mathbf{x}^{k+1}$

$$\mathbf{x}^{k+1} = \left[\beta\sum_q \mathbf{B}^{k^t}_q(\mathbf{s}_q^k)\mathbf{B}_q^k(\mathbf{s}_q^k) + \alpha\sum_{d\in\Delta} \mathbf{F}_d^t\boldsymbol{\Omega}_d^k\mathbf{F}_d \right]^{-1} \sum_q \mathbf{B}_q^k(\mathbf{s}_q^k)^t \left[\beta\mathbf{W}\mathbf{a}_q^k - \lambda_q^k\right] \tag{62}$$

where

$$\Omega_d^k(i,i) = \min(1/|\omega_d^{\mathbf{x}^k}(i)|^2, 1/\epsilon^2). \tag{63}$$

3.2. Transformed Coefficient Estimation

The optimisation step in Equation (51) for each $\mathbf{a}_q$ produces

$$\mathbf{a}_q^{k+1} = \arg\min_{\mathbf{a}_q} \left\{ \frac{\eta}{2} \| \boldsymbol{\Phi}\mathbf{W}\mathbf{a}_q - \mathbf{y}_q \|^2 + \tau \| \mathbf{a}_q \|_1 + \frac{\beta}{2} \| \mathbf{B}_q^k(\mathbf{s}_q^k)\mathbf{x}^{k+1} - \mathbf{W}\mathbf{a}_q \|^2 - \boldsymbol{\lambda}_q^{k^t}(\mathbf{B}_q^k(\mathbf{s}_q^k)\mathbf{x}^{k+1} - \mathbf{W}\mathbf{a}_q) \right\}, \tag{64}$$

which is equivalent to

$$\mathbf{a}_q^{k+1} = \arg\min_{\mathbf{a}_q} \left\{ \frac{\eta}{2} \parallel \mathbf{\Phi W a}_q - \mathbf{y}_q \parallel^2 \right.$$
$$\left. + \frac{\beta}{2} \parallel \mathbf{B}_q^k(\mathbf{s}_q^k)\mathbf{x}^{k+1} - \boldsymbol{\lambda}_q^k - \mathbf{W a}_q \parallel^2 + \tau \parallel \mathbf{a}_q \parallel_1 \right\}. \tag{65}$$

The above equation can be rewritten as

$$\mathbf{a}_q^{k+1} = \arg\min_{\mathbf{a}_q} \parallel \mathbf{\Phi}' \mathbf{W a}_q - \mathbf{J}' \parallel^2 + \tau \parallel \mathbf{a}_q \parallel_1 , \tag{66}$$

where

$$\mathbf{J}' = \begin{bmatrix} \sqrt{\frac{\eta}{2}} \mathbf{y}_q \\ \sqrt{\frac{\beta}{2}} (\mathbf{B}_q^k(\mathbf{s}_q^k)\mathbf{x}^{k+1} - \boldsymbol{\lambda}_q^k) \end{bmatrix} \text{ and } \mathbf{\Phi}' = \begin{bmatrix} \sqrt{\frac{\eta}{2}} \mathbf{\Phi} \\ \sqrt{\frac{\beta}{2}} \mathbf{I} \end{bmatrix} \tag{67}$$

with $\mathbf{I}$ the $D \times D$ identity matrix. The optimisation problem in Equation (66) can be solved using the algorithm in [Kim et al., 2007].

3.3. Registration from Estimated HR Image

To estimate the registration parameters in Equation (52), we solve

$$\mathbf{s}_q^{k+1} = \arg\min_{\mathbf{s}_q} \frac{\beta'}{2} \parallel \mathbf{B}_q(\mathbf{s}_q)\mathbf{x}^{k+1} - \mathbf{W a}_q^{k+1} \parallel^2, \tag{68}$$

where β' is a positive parameter. This has been approached differently in [Saafin et al., 2015a, Saafin et al., 2015b]. However the next two approaches reported better results.

$\mathbf{B}_q(\mathbf{s}_q)\mathbf{x}$, can be approximated by expanding it into its first-order Taylor series around the previous value $\mathbf{s}_q^k$. Hence obtaining (see [He et al., 2007, Villena et al., 2013])

$$\mathbf{B}_q(\mathbf{s}_q)\mathbf{x}^{k+1} \approx \mathbf{B}_q(\mathbf{s}_q^k)\mathbf{x}^{k+1}$$
$$+ \left[\mathbf{O}_{q1}(\mathbf{s}_q^k)\mathbf{x}^{k+1}, \mathbf{O}_{q2}(\mathbf{s}_q^k)\mathbf{x}^{k+1}, \mathbf{O}_{q3}(\mathbf{s}_q^k)\mathbf{x}^{k+1} \right] (\mathbf{s}_q - \mathbf{s}_q^k), \tag{69}$$

where $\mathbf{O}_{qi}(\mathbf{s}_q^k)\mathbf{x}^{k+1} = \mathbf{AH}_q\mathbf{N}_i(\mathbf{s}_q^k)\mathbf{x}^{k+1}$, and $\mathbf{N}_i(\mathbf{s}_q^k)\mathbf{x}^{k+1}$ is defined as

$$\begin{aligned}&\left[\mathbf{N}_1(\mathbf{s}_q^k)\mathbf{x}^{k+1}, \mathbf{N}_2(\mathbf{s}_q^k)\mathbf{x}^{k+1}, \mathbf{N}_3(\mathbf{s}_q^k)\mathbf{x}^{k+1}\right] = \\ &\left[(\mathbf{P}_1(\mathbf{s}_q^k)\mathbf{M}_1(\mathbf{s}_q^k) + \mathbf{P}_2(\mathbf{s}_q^k)\mathbf{M}_2(\mathbf{s}_q^k), \mathbf{M}_1(\mathbf{s}_q^k), \mathbf{M}_2(\mathbf{s}_q^k)\right].\end{aligned} \tag{70}$$

In Equation (70)

$$\mathbf{M}_1(\mathbf{s}_q^k) = (\mathbf{I} - \mathbf{D}_{\mathbf{b}_q(\mathbf{s}_q)})(\mathbf{L}_{tr(\mathbf{s}_q)} - \mathbf{L}_{tl(\mathbf{s}_q)}) + \mathbf{D}_{\mathbf{b}_q(\mathbf{s}_q)}(\mathbf{L}_{br(\mathbf{s}_q)} - \mathbf{L}_{bl(\mathbf{s}_q)}) \tag{71}$$

$$\mathbf{M}_2(\mathbf{s}_q^k) = (\mathbf{I} - \mathbf{D}_{\mathbf{a}_q(\mathbf{s}_q)})(\mathbf{L}_{bl(\mathbf{s}_q)} - \mathbf{L}_{tl(\mathbf{s}_q)}) + \mathbf{D}_{\mathbf{a}_q(\mathbf{s}_q)}(\mathbf{L}_{br(\mathbf{s}_q)} - \mathbf{L}_{tr(\mathbf{s}_q)}) \tag{72}$$

$$\mathbf{P}_1(\mathbf{s}_q^k) = -[\mathbf{D}_\mathbf{u}\sin(\theta_q^k) + \mathbf{D}_\mathbf{v}\cos(\theta_q^k)] \tag{73}$$

$$\mathbf{P}_2(\mathbf{s}_q^k) = [\mathbf{D}_\mathbf{u}\cos(\theta_q^k) - \mathbf{D}_\mathbf{v}\sin(\theta_q^k)]. \tag{74}$$

and $\mathbf{D_u}$ and $\mathbf{D_v}$ are diagonal matrices whose diagonals respectively are the vectors $\mathbf{u}$ and $\mathbf{v}$, representing pixel coordinates in $\mathbf{x}$. Substituting Equation (69) into Equation (68), we obtain the following update equation

$$\mathbf{s}_q^{k+1} = \left[\mathbf{\Lambda}_q^k\right]^{-1}(\Upsilon_q^k + \mathbf{\Lambda}_q^k\mathbf{s}_q^k) = \mathbf{s}_q^k + \left[\mathbf{\Lambda}_q^k\right]^{-1}\Upsilon_q^k, \tag{75}$$

where $\mathbf{\Lambda}_q^k$ and Υ_q^k correspond to the $q-$th observation at the $k-$th iteration with respectively $(i, j) \in \{1, 2, 3\}$ element and $i \in \{1, 2, 3\}$ element given by

$$\mathbf{\Lambda}_{qij}^k = \left[\mathbf{AH}_q\mathbf{N}_i(\mathbf{s}_q^k)\mathbf{x}^{k+1}\right]^t \mathbf{AH}_q\mathbf{N}_j(\mathbf{s}_q^k)\mathbf{x}^{k+1}, \tag{76}$$

$$\Upsilon_{qi}^k = \left[\mathbf{AH}_q\mathbf{N}_i(\mathbf{s}_q^k)\mathbf{x}^{k+1}\right]^t (\mathbf{Wa}_q^{k+1} - \mathbf{AH}_q\mathbf{N}_i(\mathbf{s}_q^k)\mathbf{x}^{k+1}). \tag{77}$$

where $\mathbf{N}_i$ were defined in Equation (70).

3.4. Registration from Estimated LR Reference Image

In Section 4.6 the registration estimation procedure using Equation (68) will be compared with registration from the LR observations using the following equation,

$$\mathbf{s}_{Lq}^{k+1} = \arg\min_{\mathbf{s}_{Lq}} \frac{\beta'}{2} \parallel \mathbf{C}'(\mathbf{s}_{Lq})\mathbf{z}_r^{k+1} - \mathbf{z}_q^{k+1} \parallel^2. \tag{78}$$

where $\mathbf{C}'(\mathbf{s}_{Lq})$ is a $D \times D$ warping matrix, $\mathbf{z}_r^{k+1} = \mathbf{W}\mathbf{a}_r^{k+1}$ is the LR image used for registration (reference image), and $\mathbf{z}_q^{k+1} = \mathbf{W}\mathbf{a}_q^{k+1}$ is the $q-$th LR observation, obtained in the previous step of the iterative process. Notice that the warped LR observation $\mathbf{C}'(\mathbf{s}_{Lq})\mathbf{z}_r$ is equivalent to the LR version of $\mathbf{C}(\mathbf{s}_q)\mathbf{x}$.

By analogy, the following update equation can be obtained

$$\mathbf{s}_{Lq}^{k+1} = \mathbf{s}_{Lq}^{k} + \left[\mathbf{\Lambda}_{Lq}^{k}\right]^{-1} \Upsilon_{Lq}^{k}, \tag{79}$$

where $\mathbf{\Lambda}_{Lq}^{k}$ and Υ_{Lq}^{k} correspond to the $q-$th observation at the $k-$th iteration with respectively $(i,j) \in \{1,2,3\}$ element and $i \in \{1,2,3\}$ element given by

$$\Lambda_{Lqij}^{k} = \left[\mathbf{N}_i(\mathbf{s}_{Lq}^{k})\mathbf{z}_r^{k+1}\right]^t \mathbf{N}_j(\mathbf{s}_{Lq}^{k})\mathbf{z}_r^{k+1}, \tag{80}$$

$$\Upsilon_{Lqi}^{k} = \left[\mathbf{N}_i(\mathbf{s}_{Lq}^{k})\mathbf{z}_r^{k+1}\right]^t \left(\mathbf{z}_q^{k+1} - \mathbf{N}_i(\mathbf{s}_{Lq}^{k})\mathbf{z}_r^{k+1}\right), \tag{81}$$

where $\mathbf{N}_i(\mathbf{s}_{Lq}^{k})\mathbf{z}_r^{k+1}$ is defined as

$$\begin{aligned}&\left[\mathbf{N}_1(\mathbf{s}_{Lq}^{k})\mathbf{z}_r^{k+1}, \mathbf{N}_2(\mathbf{s}_{Lq}^{k})\mathbf{z}_r^{k+1}, \mathbf{N}_3(\mathbf{s}_{Lq}^{k})\mathbf{z}_r^{k+1}\right]\\ &\quad= \left[(\mathbf{P}_1(\mathbf{s}_{Lq}^{k})\mathbf{M}_1(\mathbf{s}_{Lq}^{k}) + \mathbf{P}_2(\mathbf{s}_{Lq}^{k})\mathbf{M}_2(\mathbf{s}_{Lq}^{k}), \mathbf{M}_1(\mathbf{s}_{Lq}^{k}), \mathbf{M}_2(\mathbf{s}_{Lq}^{k})\right],\end{aligned} \tag{82}$$

If $\mathbf{s}_{Lq}^{k+1} = [\theta_{Lq}, c_{Lq}, d_{Lq}]^t$ then the HR motion vector is

$$\mathbf{s}_q^{k+1} = [\theta_{Lq}, P \cdot c_{Lq}, P \cdot d_{Lq}]^t. \tag{83}$$

3.5. CSSR Algorithm Statement

The complete CSSR algorithm is presented in Algorithm 1. Notice here the two possible registration procedures from the HR image using Equation (75) and from the LR reference image using Equation (83).

4. Experimental Results

The CSSR reconstruction problem estimates one HR image of a scene. Usually, we do not have the original image to compare with the estimated one. However, there are some measures that can help indicating the performance level of the estimation process.

Algorithm 1 Compressive Sensing Super Resolution (CSSR) Algorithm

Require: Values α, β, τ, η
Initialize $\mathbf{a}^0$, $\mathbf{s}^0$, $\boldsymbol{\lambda}^0$, $\boldsymbol{\Omega}^0 = \{\boldsymbol{\Omega}_d^0, d \in \Delta\}$
$k = 0$
while convergence criterion is not met **do**
1. Calculate $\mathbf{x}^{k+1}$ by solving Equation (62)
2. For $d \in \Delta$, calculate $\boldsymbol{\Omega}_d^{k+1}$ using Equation (63)
3. For $q = 1, \ldots, Q$, calculate $\mathbf{a}_q^{k+1}$ using Equation (66)
4. For $q = 1, \ldots, Q$, calculate $\mathbf{s}_q^{k+1}$ using one of Equations (75) or (83)
5. For $q = 1, \ldots, Q$, update $\boldsymbol{\lambda}_q^{k+1}$ using Equation (53)
6. Set $k = k + 1$
end while
return $\mathbf{x} = \mathbf{x}^k$

The first assessment criterion is the human eye perception which to some limit can be adopted. However, it does not give a standard measure and it may depend on the person himself looking at the estimated image.

Another assessment criteria which may give a better judgement are machine vision techniques, where the estimated image is forwarded to another processing step to make a decision or to operate a device. The reliability and accuracy measures of the machine are known, then the merits of the estimated image, which is fed as the input to the machine, will affect the performance of the machine. The change in overall performance of the machine definitely gives an indication about how much efficient the estimation process was. As an example, let us consider a set of PMMW images, how to measure the performance of the super resolved image? The answer is that when the HR image is fed to a threat detection algorithm, and this leads to a higher reliability then this will be a consequence of the higher quality of the estimated image.

To simplify matters, usually input images can be simulated to approximate the real ones. If the original image is $\mathbf{x}$, then we can generate LR versions of it. Then, these simulated input images are fed to the algorithm, to find an estimate of the HR image $\hat{\mathbf{x}}$. Notice that by this we already have both the original image and the estimated image. In such a case, mathematical measures can be used to assess the performance of the proposed algorithm.

Many performance measures have been utilised in image processing, like the Mean Square Error (MSE) and Peak Signal to Noise Ratio (PSNR), explained next.

MSE is the average squared difference error between the estimated and original images. It is calculated by

$$MSE = \frac{1}{N} \parallel \mathbf{x} - \hat{\mathbf{x}} \parallel^2 , \tag{84}$$

where $\mathbf{x}$ and $\hat{\mathbf{x}}$ are $N \times 1$ vectors, representing respectively the original and estimated images, and $\parallel \cdot \parallel$ is the Euclidean norm.

PSNR measures the ratio between the maximum possible power in the image to the power of the fidelity term, and it is expressed as

$$PSNR = 10 \ \log \frac{\max(\mathbf{x})^2}{MSE} , \tag{85}$$

where $\max(\mathbf{x})$ is the maximum possible value of the image $\mathbf{x}$. Notice here that $\max(\cdot)$ is not a function of the information in the image, instead it is a function of how the image is represented. If the image intensities are saved using 8-bit integer representation then the maximum possible value is 255. If the image is saved in a normalised format then the maximum value is 1.0.

Both measures were used in the simulated experiments to indicate the quality of output image. To define a robust stopping criterion for the iterative process the following condition is used

$$\frac{\parallel \hat{\mathbf{x}}^k - \hat{\mathbf{x}}^{k-1} \parallel}{\parallel \hat{\mathbf{x}}^{k-1} \parallel} \leq 10^{-3} , \tag{86}$$

where $\hat{\mathbf{x}}^k$ and $\hat{\mathbf{x}}^{k-1}$ are the estimated images at iterations k and $k-1$, respectively.

4.1. CSSR vs SR Works

This section compares the performance of the CSSR algorithm with the Bicubic Interpolation (BIC) and a set of existing SR algorithms, namely, SR using an l_1 prior [Villena et al., 2009] (L1S), SR using SAR priors [Villena et al., 2013] (SAS), a fast and robust SR [Farsiu et al., 2004] (FRSR), and a robust SR method [Zomet et al., 2001] (RSR). Algorithms L1S [Villena et al., 2009] and SAS [Villena et al., 2013] estimate all their parameters. Since algorithms FRSR [Farsiu et al., 2004] and RSR [Zomet et al., 2001] need some parameters to be set, we performed an exhaustive search to find the parameters resulting

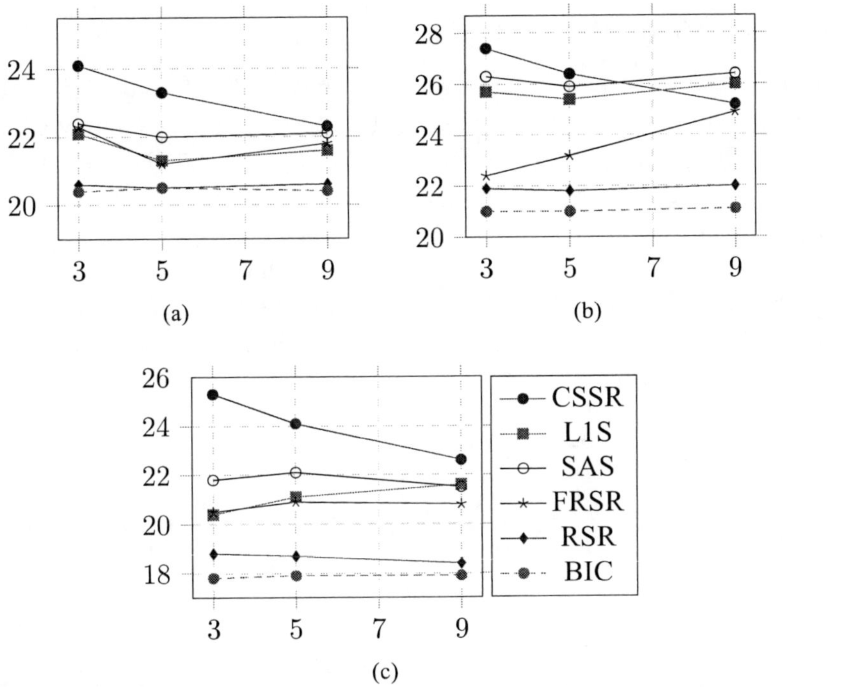

Figure 8. Comparison between SR algorithms and the CSSR algorithm (PSNR vs BlurVar). P=4, SNR=30dB, Q=4, and for CSSR, R=1.0. (a) Cameraman image, (b) Lena, (c) Shepp-Logan image.

in the maximum PSNR values. We used the Cameraman, Shepp-Logan and Lena images. The number of observations was Q=4. For the CSSR, we used a unity compression ratio R=1.0 to compare with those SR algorithms which do not use CS observations. Notice that we still use CS but with 100% of the measurements which is equivalent to using $\mathbf{\Phi} = \mathbf{I}$.

The plots in Figure 8 compare the performance measure (PSNR) of SR algorithms as a function of the variance of the blur affecting the test images for SNR=30 dB, Q=4, and P=4. Notice the relative improvement of the CSSR algorithm, when compared with SR algorithms that do not use compressed sensing.

4.2. CSSR vs ID Works

In this section the performance of the proposed CSSR algorithm is compared with a set of existing image deconvolution CS algorithms, namely L1-LS [Kim et al., 2007], GPSR [Figueiredo et al., 2007], NESTA [Becker et al., 2011], YALL1 [Yang and Zhang, 2011], and CoSaMP [Needell and Tropp, 2009], using their native codes. These algorithms estimate $\mathbf{x}$ from one CS blurred observation. CSSR used P=1 and Q=1 to comply with these works.

Figure 9 shows plots of PSNR vs compression ratio (R), for the test images, using Blur var=5 and SNR=30 dB. It is clear the quality level the CSSR offers in comparison to other works using one input CS observation. The performance of CSSR is expected to be better if multiple input images are used. Next we

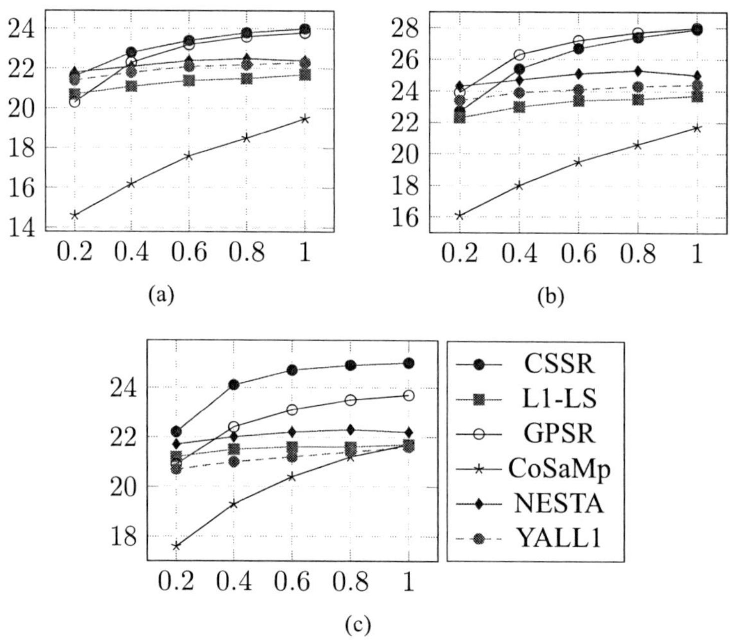

Figure 9. Comparison between CSSR and CS ID algorithms: PSNR vs R (Blur Var=5, SNR=30dB, for CSSR P=1, Q=1). (a) Cameraman, (b) Lena, (c) Shepp-Logan.

test the performance as a function of the variance of the blur. To compare the performance for all algorithms, plots are shown for the case R=0.6 and SNR=30 dB in Figure 10.

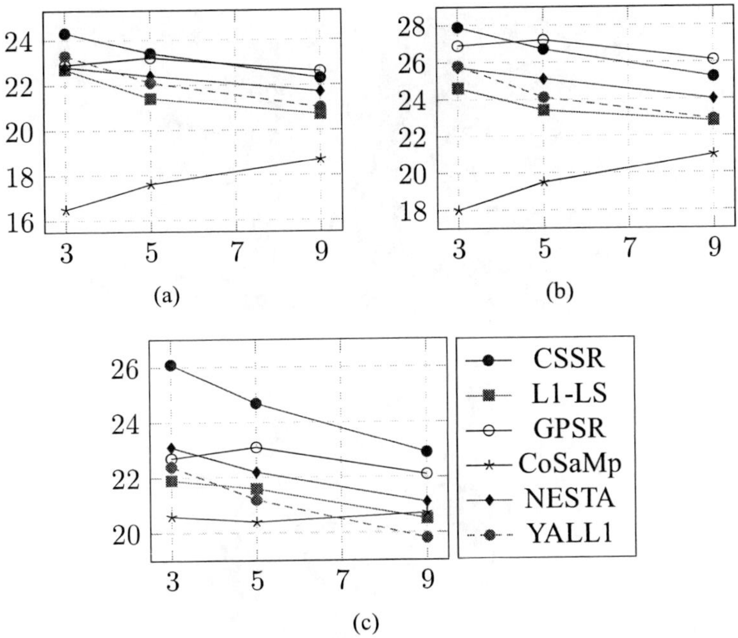

Figure 10. Comparison between CSSR and CS ID algorithms: PSNR vs BlurVar (R=0.6, SNR=30dB, for CSSR P=1, Q=1). (a) Cameraman, (b) Lena, (c) Shepp-Logan.

4.3. CSSR: The General Case

In this section the overall behavior of the CSSR algorithm is investigated; both the compression ratio and zooming factor will be varied and the results analysed. Notice here that we test the CSSR in its general case for different R and P values, while in the previous two sections R, P or Q were fixed at 1 to be able to compare with existing works which do not adapt to this general case. In this section we use the images shown in Figure 11, namely Satellite, Lena, and Peppers images.

Next we will choose four examples to precisely explore the CSSR performance. For every zooming factor, two cases were chosen: in one the image is affected by simple degradations while in the other is affected with hard degradations. These are:

- Case 1. P=2, Blur Variance=3, SNR=40 dB.
- Case 2. P=2, Blur Variance=5, SNR=30 dB.

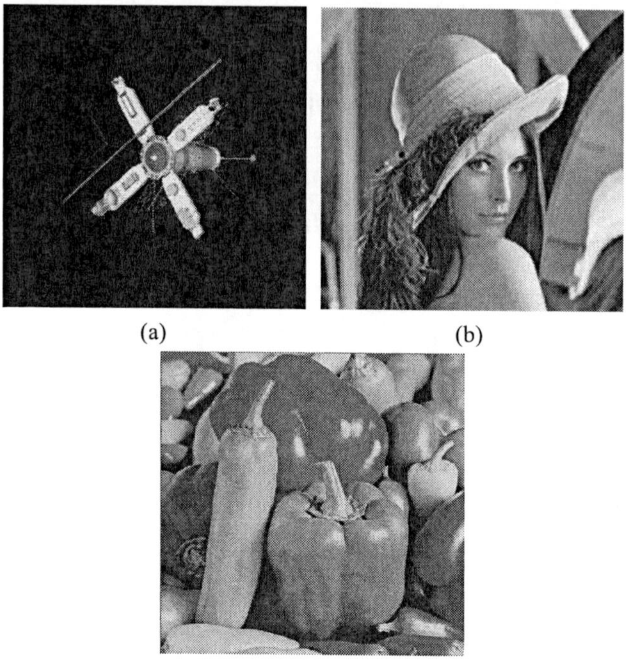

(a) (b)

(c)

Figure 11. Original Images. (a) Satellite, (b) Lena, (c) Peppers.

- Case 3. P=4, Blur Variance=3, SNR=40 dB.
- Case 4. P=4, Blur Variance=5, SNR=30 dB.

Figure 12 shows plots of the PSNR vs R. These plots, show how the PSNR, as a function of compression ratio, is affected when changing P, blur variance, and the noise added to the input images. As expected, PSNR increases when decreasing the zooming factor, decreasing the blur, increasing R, and increasing the SNR of the observed image.

Let us now visually assess the quality of the CSSR method by watching the estimated images for some examples. To do so we choose R=0.8 to show the estimated images in Figure 13. In all these experiments the number of observations was fixed at Q=5, while the values of P and the blur variance have been varied. Figures 13 to 15(a-d) used (P=2, Blur Var 3), (P=2, Blur Var=5), (P=4, Blur Var 3), and (P=4, Blur Var=5), respectively.

All the estimated images shown in Figures 13 to 15(a-d) present a good

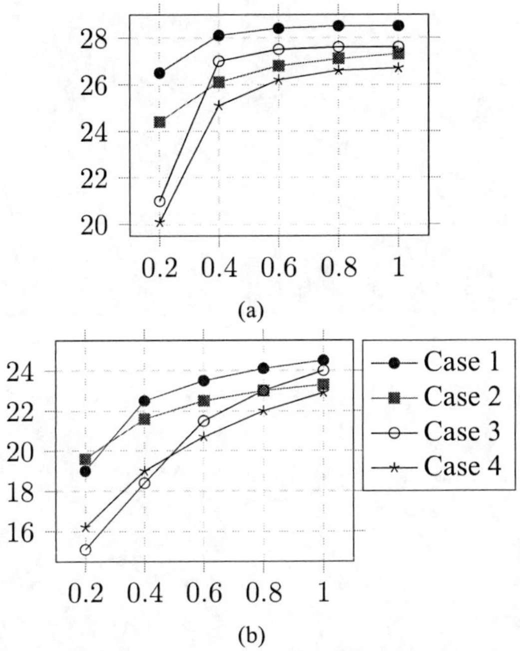

Figure 12. Performance of the CSSR algorithm (PSNR vs R) (Q=4). Case 1: P=2, Blur Var=3, SNR=40 dB, Case 2: P=2, Blur Var=5, SNR=30 dB, Case 3: P=4, Blur Var=3, SNR=40 dB, Case 4: P=4, Blur Var=5, SNR=30 dB for Satellite (a) and Barbara (b) images.

visual quality. In the next sub-section we investigate the performance of the CSSR algorithm further to the case where only one input image is available.

4.4. CSSR from a Single Image

In this section, the performance of the CSSR algorithm is investigated when the input is a single image. Although it is not the typical case, however the CSSR can be adapted for this problem. Notice here that the problem is simpler since there are no registration parameters to be estimated but the performance is not expected to be better than the case where multiple input images are available.

To compare this with the multiple input case, with Q=4, we chose P=4, Blur Var=3, SNR=40 dB for Satellite and Barbara images. The plots are shown in Figure 16. The plots show how increasing the input images can affect positively

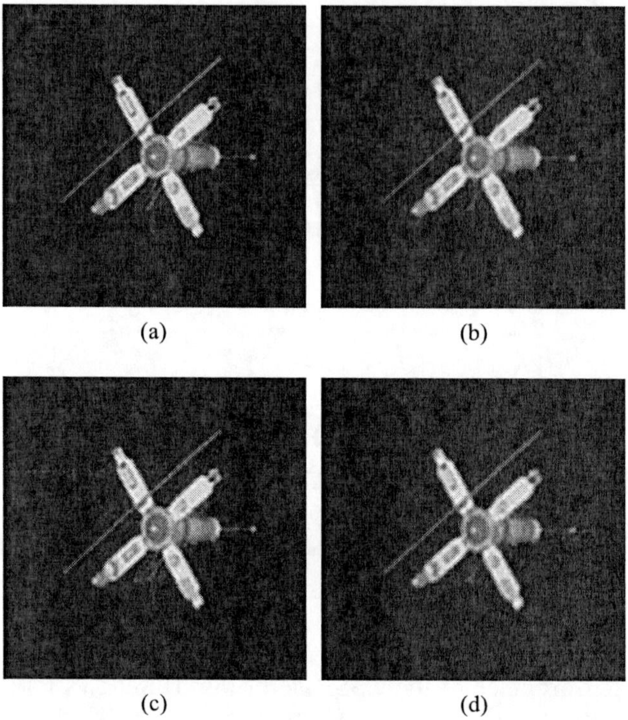

Figure 13. Estimated Satellite Images using the proposed CSSR method (R=0.8 and Q=5). (a) P=2, Blur Var=3, (b) P=2, Blur Var=5, (c) P=4, Blur Var=3, (d) P=4, Blur Var=5.

the estimated image quality. However, at very low ratios R, which are not practical to be adapted, a single image processing may, in some cases, give a better performance.

4.5. CSSR vs Sequential Approach

As mentioned earlier, the alternate approach is expected to produce a better estimated image quality. The next experiment has been executed on both the Satellite and Barbara images. We used four simulated CS LR images to test the sequential and the CSSR alternate estimations.

Figure 17 shows plots of the PSNR vs R, for the Satellite image, for the two approaches. P=4, for the two cases in Figure 17(a) Blur Var=5, SNR=30dB and

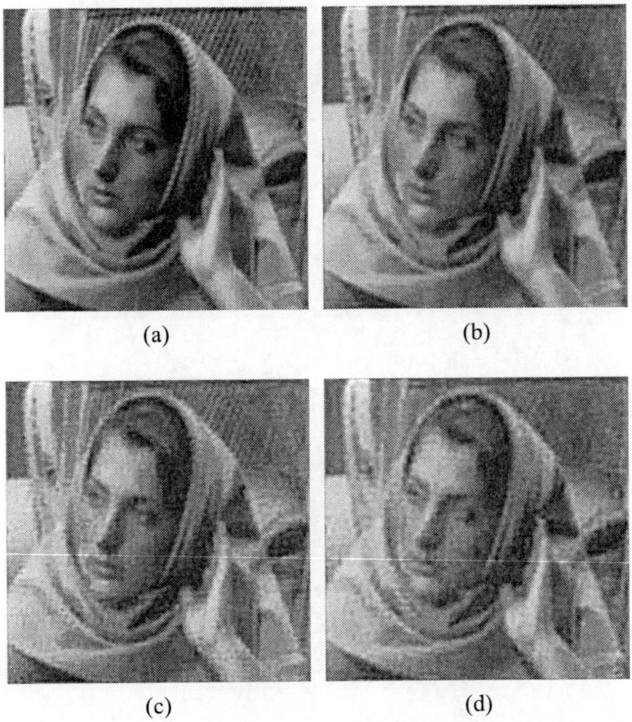

Figure 14. Estimated Barbara Images using the proposed CSSR method (R=0.8 and Q=5). (a) P=2, Blur Var=3, (b) P=2, Blur Var=5, (c) P=4, Blur Var=3, (d) P=4, Blur Var=5.

Figure 17(b) Blur Var=3, SNR=40 dB. In all cases the obtained PSNR using the alternate CSSR was better than when using the sequential approach. In the plots of the Satellite image in Figure 17(a) the alternate CSSR improves the PSNR by more than 1.2 dB when compared to the sequential approach for all Rs.

4.6. CSSR1 vs CSSR2

In this section the performance is tested for two different registration approaches. To distinguish the two used approaches, we denote by CSSR1 the one using Equation (68), and CSSR2 using Equation (78).

Figure 18 shows the plots for the Satellite image (using P=4 and Q=4). Figure 18(a) analyses the case when Blur Var=5 and SNR=30 dB, while Fig-

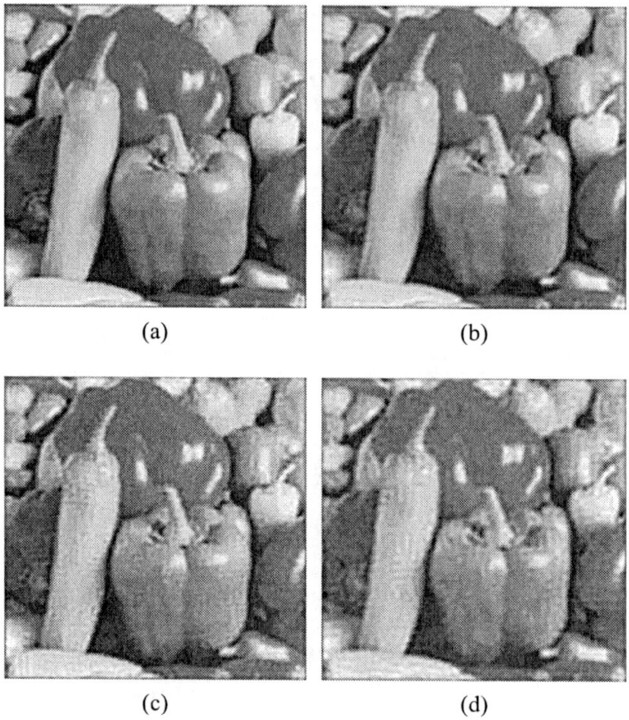

Figure 15. Estimated Peppers Images using the proposed CSSR method (R=0.8 and Q=5). (a) P=2, Blur Var=3, (b) P=2, Blur Var=5, (c) P=4, Blur Var=3, (d) P=4, Blur Var=5.

ure 18(b) analyses the case when Blur Var=3, SNR=40 dB. This proves that the two registration approaches are effective and accurate and can be adopted interchangeably. This is clear in the plots where the curves coincide. Experimental results were performed on additional simulated and real images in our published works [AlSaafin et al., 2016].

Two applications are described in the next two sections: Passive Millimeter Wave (PMMW) and colour images.

5. CSSR of PMMW Images

Due to their characteristics, MMW images are used in applications like weather operations, low visibility navigation, and the imaging of people for concealed

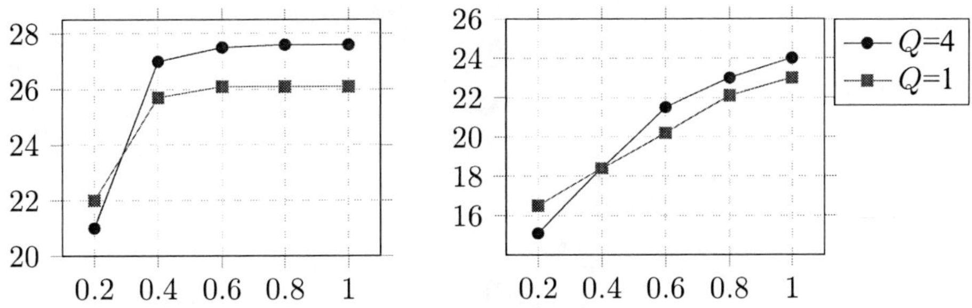

(a) (b)

Figure 16. CSSR performance from multiple and single input images. (P=4, Blur Var=3, SNR=40 dB. on (a) Satellite and (b) Barbara images.

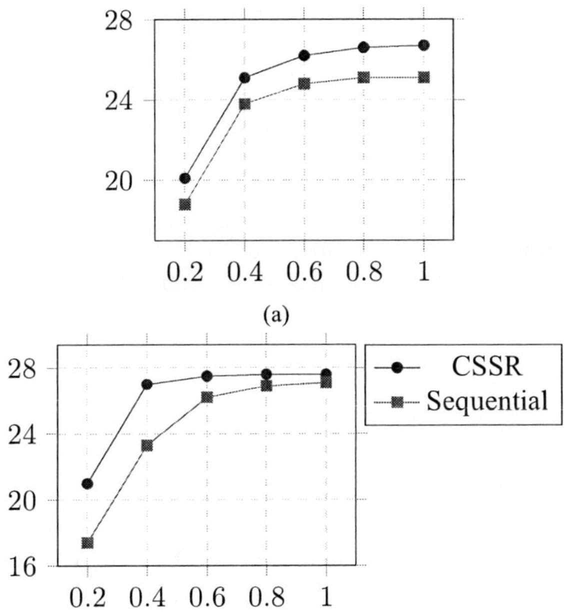

(b)

Figure 17. Comparison of the CSSR vs the sequential approach for the Satellite image, (Q=4, P=4), using (a) Blur Var=5, SNR=30 dB, and (b) Blur Var=3, SNR=40 dB.

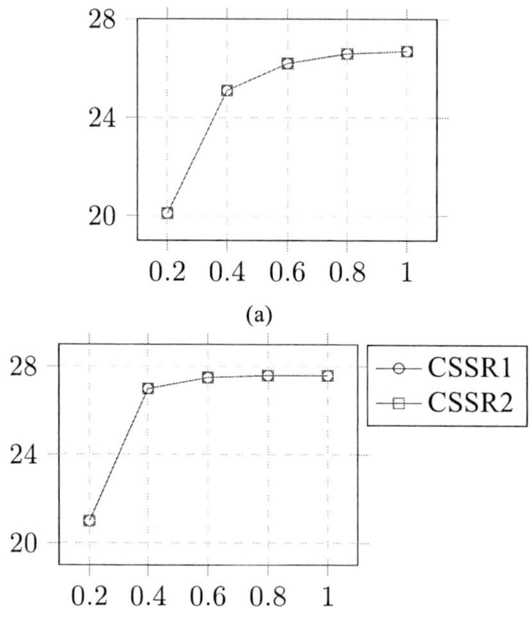

Figure 18. Comparison of the proposed CSSR1 vs CSSR2 for the Satellite image, (Q=4, P=4), using (a) Blur Var=5, SNR=30 dB, and (b) Blur Var=3, SNR=40 dB.

object and threat detection [Yujiri et al., 2003, Chen et al., 2005, Zheng et al., 2013], just to name a few.

Based on their interest, image processing techniques have recently started to be applied to these images. For instance, PMMW image enhancement has been addressed in [Yang et al., 2010a, Li et al., 2011, Chen et al., 2015, Chen et al., 2005, Lee et al., 2011] to perform detection tasks. In [Li et al., 2011] the high frequency components of those images were restored using a MAP estimator, and they were then added to the input image to produce an HR image. Registration and fusion of visible and MM images as well as segmentation of MMW images have been addressed in [Chen et al., 2005, Lee et al., 2011, Chen et al., 2015].

Unfortunately PMMW systems have two serious shortcomings: the long acquisition time needed to produce a PMMW image and the poor resolution of captured images. CS systems, with its reduced image acquisition time, have been applied to PMMW imaging, see [Chan et al., 2008, Heidari and Saeedkia, 2009, Gopalsami et al., 2012, Gopalsami et al., 2011]. In [Gopalsami et al., 2012, Gopalsami et al., 2011] the authors utilise Hadamard masks to reduce the acquisition time in PMMW imagers. In [Patel and Mait, 2012] a PMMW imaging system with extended depth-of-field that can produce images with reduced number of samples is presented. CS and BID have been combined in [Amizic et al., 2013a]. By end large the resolution of MMW images is small, therefore making their analysis a challenging task.

In [AlSaafin et al., 2016, Saafin et al., 2015b] HR images were super resolved for the first time from multiple CS observations of unregistered LR PMMW images. The PMMW images represent an important application area where CS and LR to HR techniques can be combined to enhance the performance of related application capabilities of current PMMW imaging systems.

To test the CSSR results experiments were performed on real input images, acquired by a PMMW imager. These images were synthetically compressed using a circulant Toeplitz matrix following the Bernoulli probability distribution.

Figure 19(a-d) show four real PMMW LR observations, of a man hiding a threat attached to his left arm. The estimated HR image using bilinear interpolation of one reconstructed LR image, is shown in Figure 19(e), and the estimated using the CSSR algorithm, with Q=4, is shown in Figure 19(f) (P=2, R=0.8). Also, the corresponding estimated images using P=4, R=0.8 are shown in Figure 19(g) for bilinear interpolation, and Figure 19(h) for the CSSR algorithm. The obtained images clearly show the better reconstruction capabilities of the proposed method, this can be observed by the clear discontinuity of the arm indicating the attached threat.

6. CSSR of Colour Images

There is no much work dealing with colour images from CS observations. [Majumdar and Ward, 2010] proposed a method for colour imaging via compressed sensing. It works on the colour channels and random projections from each channel, Red, Green and Blue (RGB), are acquired separately. The reconstruction process makes use of the correlation between the three colour channels and it is based on group sparse optimisation.

In [Nagesh and Li, 2009], the single-pixel CS camera has been combined with a Bayer colour filter to acquire CS color images, then joint sparsity models were applied to recover the three RGB channels. [Sen and Darabi, 2009] esti-

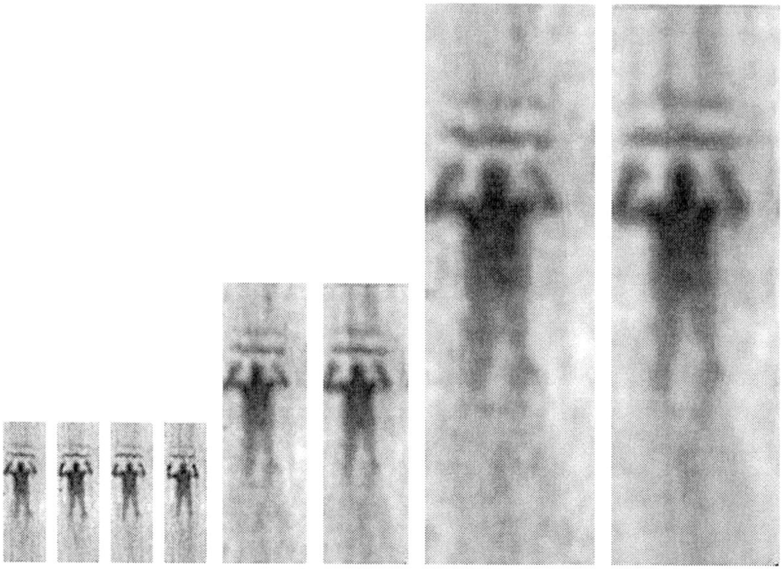

(a) (b) (c) (d) (e) (f) (g) (h)

Figure 19. PMMW image of a man with a threat attached to his arm (Q=4, R=0.8). (a-d) LR images, (e) Bilinear interpolation from one reconstructed image, P=2, (f) Estimated HR image using the CSSR algorithm, P=2, (g) Bilinear interpolation from one reconstructed image, P=4, (h) Estimated HR image using the CSSR algorithm, P=4.

mated SR images from a single LR CS observation and applied this to colour images. In [Sankaranarayanan et al., 2012] spatial multiplexing cameras were used to sample the scene through a series of coded projections, this was applied to CS video super resolution and colour videos.

To estimate the registration parameters from the estimated HR image, the following equation is solved

$$\mathbf{s}_q^{k+1} = \arg\min_{\mathbf{s}_q} \frac{\beta'}{2} \parallel \mathbf{B}_q(\mathbf{s}_q)\mathbf{Z}^{k+1} - \mathbf{z}_q^{k+1} \parallel^2 . \tag{87}$$

where β' is a positive parameter, $\mathbf{Z} = Y(\mathbf{x})$ and $\mathbf{z}_q = Y(\mathbf{W}\mathbf{a}_q)$ represent respectively the HR and LR q-th images, and $Y(\mathbf{x})$ is calculated using

$$Y(\mathbf{x}) = 0.2989\,\mathbf{x}_R + 0.5870\,\mathbf{x}_G + 0.1140\,\mathbf{x}_B, \tag{88}$$

where $\mathbf{x}_R, \mathbf{x}_G$, and $\mathbf{x}_B$ are the R, G, and B channels of $\mathbf{x}$. Once $\mathbf{\Lambda}_q^k$ and Υ_q^k are calculated by analogy to Equations (76) and (77) and using $\mathbf{Z}$ instead of $\mathbf{x}$, the registration parameters can be calculated using Equation (75).

To estimate the registration parameters from the reference LR image, we solve the following minimisation

$$\mathbf{s}_{Lq}^{k+1} = \arg\min_{\mathbf{s}_{Lq}} \frac{\beta'}{2} \parallel \mathbf{C}'(\mathbf{s}_{Lq})\mathbf{z}_r^{k+1} - \mathbf{z}_q^{k+1} \parallel^2 . \tag{89}$$

where $\mathbf{C}'(\mathbf{s}_{Lq})$ is a $D \times D$ warping matrix, $\mathbf{z} = Y(\mathbf{Wa})$, and the index r represents the LR reference image. Once $\mathbf{\Lambda}_{Lq}^k$ and Υ_{Lq}^k are calculated as in Equations (80) and (81), the update equation is obtained using Equation (83). A complete presentation of the CSSR extended to the colour images in [Saafin et al., 2016].

Next we present experimental results obtained from colour images using the CSSR algorithm, as an important application of the CSSR paradigm. The experiments were performed using both simulated and synthetically CS real colour input images. We carried out a set of experiments on the images shown in Figures 20(a-c): Lena, Peppers, and Barbara images. The images have been warped, then degraded by Gaussian blurs of different known variances and down-sampled. The LR RGB channels have been compressed separately using a circulant Toeplitz matrix following the Bernoulli probability distribution. White Gaussian noise, with SNR=40 dB, has been finally added to the compressed channels.

Figures 20(d,e) show, respectively, examples of a simulated LR and a CS-LR observation for the Barbara image using motion vector $\mathbf{s} = [.1222, -2, 3]^t$, blur variance 7, P= 2 and R=0.5.

In the first experiment Q=3 observations and a zooming factor P=1 have been used in order to compare the CSSR method with the following CS reconstruction methods: Nagesh et. al [Majumdar and Ward, 2010] (Na) , Wakin et al [Wakin et al., 2006] (Wa), Angshul L2,1 (L1), L2,0.4 (L4), and SL2,0 (SL) [Majumdar and Ward, 2010]. Figure 21 shows a comparison between the results obtained using these CS algorithms and the obtained using the CSSR algorithm for the Lena, Barbara, and Peppers images (P=1.0 and Q=3).

In the second experiment we investigate the performance of the CSSR algorithm for practical P and R values. For all images we used Gaussian blur of variance 5, SNR=40dB, and Q=3. Shown in Figure 22(a,b) are curves of the PSNR vs R for the three images used for the cases P=2 and P=4, respectively.

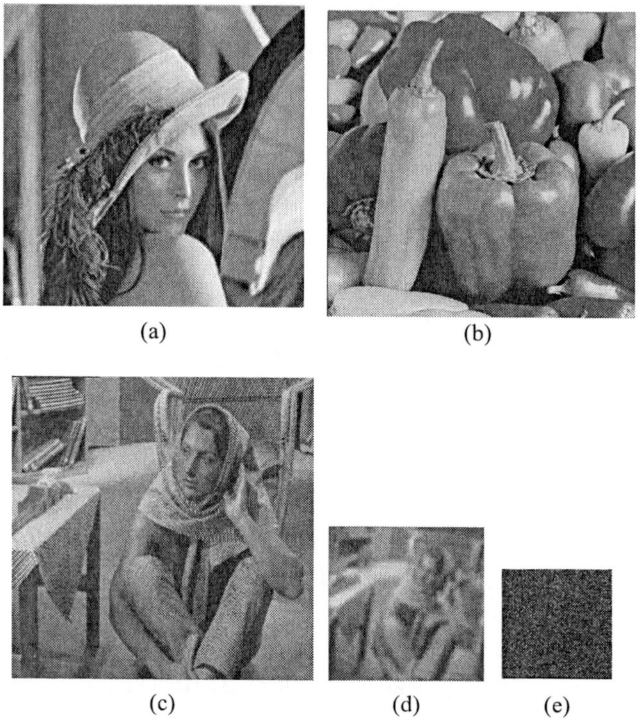

Figure 20. Original colour images (a) Lena, (b) Peppers (c) Barbara, (d) Simulated LR Barbara image ($\mathbf{s} = [.1222, -2, 3]^t$, blur variance 7, P=2), (e) simulated CS observation (R=0.5, SNR=40 dB). The black line represents the added zero-valued entries to illustrate a square image.

The reconstructed Peppers images are shown in Figures 23(a,b), using R=0.6, Blur Var=5, SNR=40dB, and Q=3, for the two cases P=2 and P=4, respectively.

The third experiment used real LR images of a toy, which have been synthetically compressed using R=0.8. We used a 16 input image sequence to reconstruct the HR image, with P=2, shown in Figure 24(d), while Figure 24(c) shows the reconstructed image obtained using only the first 4 input images of the previous sequence, shown in Figure 24(a). Figure 24(b) shows an interpolated image from the first reconstructed LR image.

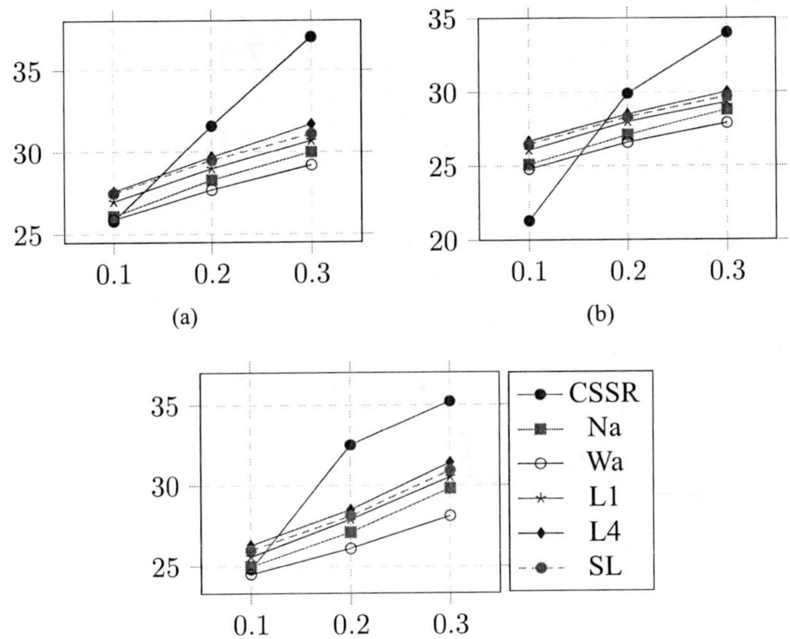

(c)

Figure 21. Comparison between CS algorithms and the CSSR algorithm for the Lena (a), Barbara (b), and Peppers (c) images. For CSSR, P=1, Q=3.

Conclusion

A novel framework to reconstruct SR images from multiple unregistered LR images acquired using CS techniques has been discussed. Any classical SR and/or CS techniques can be incorporated in the developed methodology. The degradation model of this novel combined problem has been modelled and a CSSR approach has been presented.

The CSSR method lowers the sampling frequency, transmission rates, and the bandwidth of the image signal, and memory requirements to store and/or send the image. Moreover, the acquisition time can be lowered, which is very important in medical and PMMW imaging applications. The enhanced SR image quality is advantageous in pattern recognition, medical diagnostics, and to emphasise a specific area of an image in surveillance cameras and satellite applications.

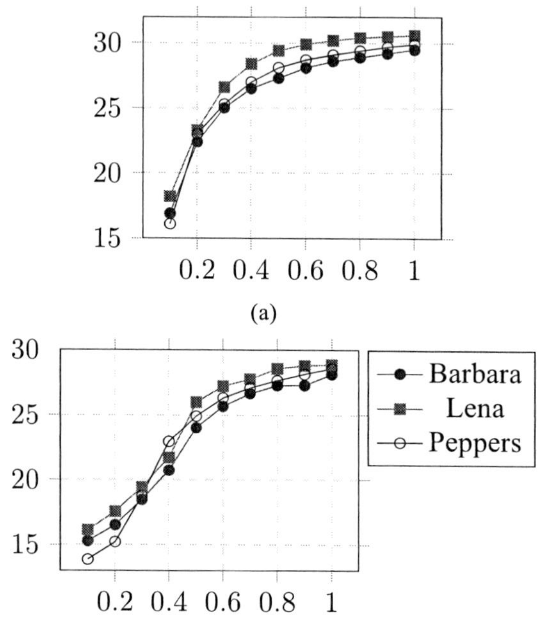

(b)

Figure 22. Performance of the proposed CSSR vs R. (Blur Var=5, SNR=40dB, Q=3). (a) P=2, (b) P=4.

The HR reconstruction problem uses a new robust sparsity promoting prior and solves a linear equation in $\mathbf{x}$ (the HR image), using conjugate gradient method. The transformed coefficient vector estimation is an ℓ_1 norm subject to a quadratic term optimisation problem. The registration parameters can be efficiently estimated using two different procedures, to finally obtain accurate results, since any SR estimation process is highly dependent on the performance of the registration estimation step. The parameters can be estimated from the estimated HR image, or the LR images themselves, with the estimation from reconstructed LR images being less sensitive to the initial conditions.

The optimisation framework uses ADMM to jointly estimate all the unknowns including HR image, LR images, and registration parameters. Experiments has shown that this simultaneous reconstruction outperforms the sequential method, that first performs LR reconstruction to then obtain an HR image from a set of LR observations. The sequential approach estimates every unknown just once, while the alternate approach iteratively updates the estimated

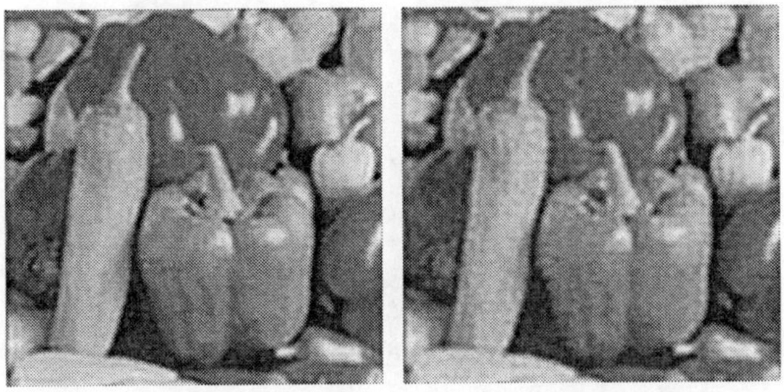

(a) (b)

Figure 23. Reconstructed Peppers colour images using the CSSR algorithm. R=0.6, Blur Var=5, SNR=40dB, and Q=3. (a) P=2, (b) P=4,

unknowns. The better performance of the alternate approach is due to the additional information provided during the estimation process.

We also showed experimentally the enhancement obtained due to the multiple input images when compared with a single one. This is expected due to the additional information every input image provides. However, the multiplicity necessitates accurate registration estimation procedure which inherently affects the overall process.

The proposed framework has been tested and compared, at a unity compression ratio, with state-of-the-art SR algorithms which do not use compression in their work. It has been also compared, at a unity zooming factor, with state-of-the-art ID algorithms that deblur CS images, which do not use down-sampling in their work. Both comparisons used synthetic images and showed better performance of the CSSR over others. The performance of CSSR when using practical values of P and R has also been tested and analysed.

Besides, the CSSR effectiveness has been demonstrated experimentally on synthetically CS noisy real images. This represents the practical application of the CSSR method, which showed very good results. The proposed framework can be extended to CS video for the estimation of both intra-frame and inter-frame SR.

The proposed CSSR method has been tested on PMMW images. These images usually have poor qualities and limited resolution, which makes super resolving them a challenging task and is a hard test of the CSSR approach. The CSSR could be used to improve the quality of PMMW images. This is expected

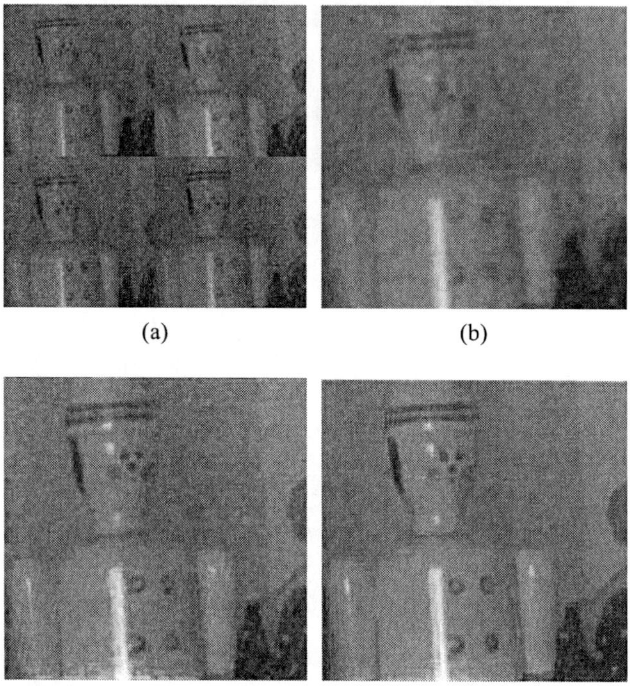

(a) (b)

(c) (d)

Figure 24. Image super resolution from real colour observations, R=0.8, P=2. (a) First 4 LR images (b) Bilinear interpolation of one reconstructed LR image, (c) Estimated HR image using the CSSR algorithm, with Q=4, (d) Estimated HR image using the CSSR algorithm, with Q=16.

to improve threat detection rates, which is an expected field of future research. Notice here that the nature of PMMW is similar to astronomical images and medical images, and the good performance of the proposed CSSR algorithm can be extended to these images.

The proposed approach could also be applied to true colour CS images. The separately sensed channels are utilised in a joint registration estimation to effectively and accurately estimate registration parameters for the three channels. The obtained results present an excellent image quality, even better than the obtained using previous CS reconstruction methods which correlate the colour channels. The efficient optimisation process can be extended in future works to deal with mosaic colour images.

References

[AlSaafin et al., 2016] AlSaafin, W., Villena, S., Vega, M., Molina, R., and Katsaggelos, A. (2016). Compressive sensing super resolution from multiple observations with application to passive millimeter wave images. *Digital Signal Processing*, pages 180–190.

[Amizic et al., 2013a] Amizic, B., Spinoulas, L., Molina, R., and Katsaggelos, A. (2013a). Compressive blind image deconvolution. *IEEE Transactions on Image Processing*, 22(10):3994–4006.

[Amizic et al., 2013b] Amizic, B., Spinoulas, L., Molina, R., and Katsaggelos, A. (2013b). Variational bayesian compressive blind image deconvolution. In *Proceedings of the 21st European Signal Processing Conference (EUSIPCO)*, pages 1–5.

[Babacan et al., 2012] Babacan, S., Molina, R., Do, M., and Katsaggelos, A. (2012). Bayesian blind deconvolution with general sparse image priors. In *Computer Vision–ECCV 2012*, pages 341–355. Springer.

[Babacan et al., 2011] Babacan, S., Molina, R., and Katsaggelos, A. (2011). Variational bayesian super resolution. *IEEE Transactions on Image Processing*, 20(4):984–999.

[Babcock et al., 2013] Babcock, H., Moffitt, J., Cao, Y., and Zhuang, X. (2013). Fast compressed sensing analysis for superresolution imaging using l1-homotopy. *Optics Express*, 21(23):28583–28596.

[Bajwa et al., 2007] Bajwa, W., Haupt, J., Raz, Gand Wright, S., and Nowak, R. (2007). Toeplitz-structured compressed sensing matrices. In *14th Workshop on Statistical Signal Processing SSP'07*, pages 294–298. IEEE.

[Becker et al., 2011] Becker, S., Bobin, J., and Candès, E. (2011). Nesta: a fast and accurate first-order method for sparse recovery. *SIAM Journal on Imaging Sciences*, 4(1):1–39.

[Belekos et al., 2010] Belekos, S., Galatsanos, N., and Katsaggelos, A. (2010). Maximum a posteriori video super-resolution using a new multichannel image prior. *IEEE Transactions on Image Processing*, 19(6):1451–1464.

[Bose et al., 2006] Bose, N., Ahuja, N., et al. (2006). Superresolution and noise filtering using moving least squares. *IEEE Transactions on Image Processing*, 15(8):2239–2248.

[Candes and Romberg, 2005] Candes, E. and Romberg, J. (2005). Practical signal recovery from random projections. *Wavelet Applications in Signal and Image Processing XI, Proc. SPIE Conf.*, 5914.

[Candes and Tao, 2005] Candes, E. and Tao, T. (2005). Decoding by linear programming. *IEEE Transactions on Information Theory*, 51(12):4203–4215.

[Candes and Wakin, 2008] Candes, E. and Wakin, M. (2008). An introduction to compressive sampling. *Signal Processing Magazine, IEEE*, 25(2):21–30.

[Capel and Zisserman, 2003] Capel, D. and Zisserman, A. (2003). Computer vision applied to super resolution. *Signal Processing Magazine, IEEE*, 20(3):75–86.

[Chambolle et al., 2010] Chambolle, A., Caselles, V., Cremers, D., Novaga, M., and Pock, T. (2010). An introduction to total variation for image analysis. *Theoretical foundations and numerical methods for sparse recovery*, 9(263-340):227.

[Chan et al., 2008] Chan, W., Charan, K., Takhar, D., Kelly, K., Baraniuk, R., and Mittleman, D. (2008). A single-pixel terahertz imaging system based on compressed sensing. *Applied Physics Letters*, 93(12):121105.

[Chen et al., 2005] Chen, H., Lee, S., Rao, R., Slamani, M., and Varshney, P. (2005). Imaging for concealed weapon detection: a tutorial overview of development in imaging sensors and processing. *IEEE Signal Processing Magazine*, 22(2):52–61.

[Chen et al., 2015] Chen, P., Zou, T., Chen, J., Gao, Z., and Xiong, J. (2015). The application of improved pso algorithm in pmmw image ostu threshold segmentation. In *Applied Mechanics and Materials*, volume 721, pages 779–782. Trans Tech Publ.

[Dai and Milenkovic, 2009] Dai, W. and Milenkovic, O. (2009). Subspace pursuit for compressive sensing signal reconstruction. *IEEE Transactions on Information Theory*, 55(5):2230–2249.

[Davenport et al., 2011] Davenport, M., Duarte, M., Eldar, Y., and Kutyniok, G. (2011). Introduction to compressed sensing. *Preprint*, 93:1–64.

[Duarte et al., 2008] Duarte, M., Davenport, M., Takhar, D., Laska, J., S, T., Kelly, K., and Baraniuk, R. (2008). Single-pixel imaging via compressive sampling. *Signal Processing Magazine, IEEE*, 25(2):83–91.

[Edeler et al., 2011] Edeler, T., Ohliger, K., Hussmann, S., and Mertins, A. (2011). Multi image super resolution using compressed sensing. In *IEEE International Conference on Acoustics, Speech and Signal Processing (ICASSP)*, pages 2868–2871.

[El-Khamy et al., 2005] El-Khamy, S., Hadhoud, M., Dessouky, M., Salam, B., and El-Samie, F. (2005). Regularized super-resolution reconstruction of images using wavelet fusion. *Optical Engineering*, 44(9):097001–097001.

[Fang et al., 2014] Fang, J., Li, J., Shen, Y., Li, H., and Li, S. (2014). Super-resolution compressed sensing: An iterative reweighted algorithm for joint parameter learning and sparse signal recovery. *IEEE Signal Processing Letters*, 21(6):761–765.

[Fannjiang and Liao, 2012] Fannjiang, A. and Liao, W. (2012). Super-resolution by compressive sensing algorithms. *CoRR*, abs/1211.5870.

[Farooque and Rohankar, 2013] Farooque, M. and Rohankar, J. (2013). Survey on various noises and techniques for denoising the color image. *International Journal of Application or Innovation in Engineering & Management (IJAIEM)*, 2(11):217–221.

[Farsiu et al., 2004] Farsiu, S., Robinson, M., Elad, M., and Milanfar, P. (2004). Fast and robust multiframe super resolution. *IEEE Transactions on Image Processing*, 13(10):1327–1344.

[Figueiredo et al., 2007] Figueiredo, M., Nowak, R., and Wright, S. (2007). Gradient projection for sparse reconstruction: Application to compressed sensing and other inverse problems. *IEEE Journal of Selected Topics in Signal Processing*, 1(4):586–597.

[Fornasier and Rauhut, 2011] Fornasier, M. and Rauhut, H. (2011). Compressive sensing. In *Handbook of mathematical methods in imaging*, pages 187–228. Springer.

[Gopalsami et al., 2011] Gopalsami, N., Elmer, T., Liao, S., Ahern, R., Heifetz, A., Raptis, A., Luessi, M., Babacan, D., and Katsaggelos, A. (2011). Compressive sampling in passive millimeter-wave imaging. In *SPIE Defense, Security, and Sensing*, pages 80220I–80220I. International Society for Optics and Photonics.

[Gopalsami et al., 2012] Gopalsami, N., Liao, S., Elmer, T., Koehl, E. R., Heifetz, A., Raptis, A. C., Spinoulas, L., and Katsaggelos, A. K. (2012). Passive millimeter-wave imaging with compressive sensing. *OPTICAL ENGINEERING*, 51(9).

[He et al., 2012] He, C., Liu, L., Xu, L., Liu, M., and Liao, M. (2012). Learning based compressed sensing for sar image super-resolution. *IEEE Journal of Selected Topics in Applied Earth Observations and Remote Sensing*, 5(4):1272–1281.

[He et al., 2007] He, Y., Yap, K., Chen, L., and Chau, L. (2007). A nonlinear least square technique for simultaneous image registration and super-resolution. *IEEE Transactions on Image Processing*, 16(11):2830–2841.

[Heidari and Saeedkia, 2009] Heidari, A. and Saeedkia, D. (2009). A 2d camera design with a single-pixel detector. In *34th International Conference on Infrared, Millimeter, and Terahertz Waves, (IRMMW-THz)*, pages 1–2.

[Huang and Makur, 2011] Huang, H. and Makur, A. (2011). Backtracking-based matching pursuit method for sparse signal reconstruction. *IEEE Signal Processing Letters*, 18(7):391–394.

[Jeengar et al., 2012] Jeengar, V., Omkar, S., Singh, A., Yadav, M., and Keshri, S. (2012). A review comparison of wavelet and cosine image transforms. *International Journal of Image, Graphics and Signal Processing (IJIGSP)*, 4(11):16.

[Ji and Fermuller, 2009] Ji, H. and Fermuller, C. (2009). Robust wavelet-based super-resolution reconstruction: theory and algorithm. *IEEE Transactions on Pattern Analysis and Machine Intelligence*, 31(4):649–660.

[Katsaggelos et al., 2007] Katsaggelos, A., Molina, R., and Mateos, J. (2007). Super resolution of images and video. *Synthesis Lectures on Image, Video, and Multimedia Processing*, 1(1):1–134.

[Kim and Kwon, 2010] Kim, K. and Kwon, Y. (2010). Single-image super-resolution using sparse regression and natural image prior. *IEEE Transactions on Pattern Analysis and Machine Intelligence*, 32(6):1127–1133.

[Kim et al., 2007] Kim, S., Koh, K., Lustig, M., Boyd, S., and Gorinevsky, D. (2007). An interior-point method for large-scale l1-regularized least squares. *IEEE Journal of Selected Topics in Signal Processing*, 1(4):606–617.

[Koh et al., 2007] Koh, K., Kim, S. J., and Boyd, S. P. (2007). An interior-point method for large-scale l1-regularized logistic regression. *Journal of Machine learning research*, 8(8):1519–1555.

[Lange, 2013] Lange, K. (2013). *Optimization.* Springer-Verlag.

[Lee et al., 2011] Lee, H., Lee, D., Yeom, S., Son, J., Guschin, V., and Kim, S. (2011). Passive millimeter wave imaging and analysis for concealed object detection. In *3rd International Conference on Data Mining and Intelligent Information Technology Applications (ICMiA)*, pages 98–101. IEEE.

[Li et al., 2013] Li, T., Wang, X., Wang, W., and Katsaggelos, A. (2013). Compressive video sensing with limited measurements. *Journal of Electronic Imaging*, 22(4):043003–043003.

[Li et al., 2011] Li, Y., Li, Y., Chen, J., and Hou, Y. (2011). Passive millimeter-wave image restoration based on improved algorithm of nonlinear extrapolation in frequency space. *International Journal of Digital Content Technology and its Applications*, 5(5):42–49.

[Majumdar and Ward, 2010] Majumdar, A. and Ward, R. (2010). Compressed sensing of color images. *Signal Processing*, 90(12):3122–3127.

[Marquina and Osher, 2008] Marquina, A. and Osher, S. (2008). Image super-resolution by tv-regularization and bregman iteration. *Journal of Scientific Computing*, 37(3):367–382.

[Meenakshi, 2015] Meenakshi, S. (2015). A survey of compressive sensing based greedy pursuit reconstruction algorithms.

[Nagesh and Li, 2009] Nagesh, P. and Li, B. (2009). Compressive imaging of color images. In *IEEE International Conference on Acoustics, Speech and Signal Processing (ICASSP)*, pages 1261–1264. IEEE.

[Nasrollahi and Moeslund, 2014] Nasrollahi, K. and Moeslund, T. (2014). Super-resolution: a comprehensive survey. *Machine vision and applications*, 25(6):1423–1468.

[Needell and Tropp, 2009] Needell, D. and Tropp, J. (2009). Cosamp: Iterative signal recovery from incomplete and inaccurate samples. *Applied and Computational Harmonic Analysis*, 26(3):301–321.

[Needell and Vershynin, 2010] Needell, D. and Vershynin, R. (2010). Signal recovery from incomplete and inaccurate measurements via regularized orthogonal matching pursuit. *IEEE Journal of Selected Topics in Signal Processing*, 4(2):310–316.

[Nelson et al., 2012] Nelson, K., Bhatti, A., and Nahavandi, S. (2012). Performance evaluation of multi-frame super-resolution algorithms. In *2012 International Conference on Digital Image Computing Techniques and Applications (DICTA)*, pages 1–8. IEEE.

[Panda et al., 2015] Panda, S., Jena, G., and Sahu, S. (2015). Image super resolution reconstruction using iterative adaptive regularization method and genetic algorithm. In *Computational Intelligence in Data Mining-Volume 2*, pages 675–681. Springer.

[Patel and Mait, 2012] Patel, V. and Mait, J. (2012). Passive millimeter-wave imaging with extended depth of field and sparse data. In *IEEE International Conference on Acoustics, Speech and Signal Processing (ICASSP)*, pages 2521–2524. IEEE.

[Pickup et al., 2009] Pickup, L., Capel, D., Roberts, S., and Zisserman, A. (2009). Bayesian methods for image super-resolution. *The Computer Journal*, 52(1):101–113.

[Rockafellar, 1997] Rockafellar, R. (1997). *Convex analysis*. Number 28. Princeton university press.

[Saafin et al., 2017] Saafin, W., Schaefer, G., Vega, M., Molina, R., and Katsaggelos, A. (2017). Image fusion of unregistered colour digital pathology images. In *Proceedings of Irish Machine Vision and Image Processing Conference (IMVIP)*.

[Saafin et al., 2015a] Saafin, W., Vega, M., Molina, R., and Katsaggelos, A. (2015a). Image super-resolution from compressed sensing observations. In *IEEE International Conference on Image Processing (ICIP)*, pages 4268–4272.

[Saafin et al., 2016] Saafin, W., Vega, M., Molina, R., and Katsaggelos, A. (2016). Compressed sensing super resolution of color images. In *Proceedings of 24th European Signal Processing Conference (EUSIPCO)*.

[Saafin et al., 2015b] Saafin, W., Villena, S., Vega, M., Molina, R., and Katsaggelos, A. (2015b). Pmmw image super resolution from compressed sensing observations. In *Proceedings of 23rd European Signal Processing Conference (EUSIPCO)*, pages 1815–1819.

[Sankaranarayanan et al., 2012] Sankaranarayanan, A., Studer, C., and Baraniuk, R. (2012). Cs-muvi: Video compressive sensing for spatial-multiplexing cameras. In *IEEE International Conference on Computational Photography (ICCP)*, pages 1–10.

[Sauer et al., 1995] Sauer, K., Borman, S., Bouman, C., et al. (1995). Parallel computation of sequential pixel updates in statistical tomographic reconstruction. In *Proceedings of International Conference on Image Processing,*, volume 2, pages 93–96. IEEE.

[Schultz and Stevenson, 1996] Schultz, R. R. and Stevenson, R. L. (1996). Extraction of high-resolution frames from video sequences. *IEEE Transactions on Image Processing*, 5(6):996–1011.

[Sen and Darabi, 2009] Sen, P. and Darabi, S. (2009). Compressive image super-resolution. In *Conference Record of the Forty-Third Asilomar Conference on Signals, Systems and Computers*, pages 1235–1242.

[Spinoulas et al., 2012] Spinoulas, L., Amizic, B., Vega, M., Molina, R., and Katsaggelos, A. (2012). Simultaneous bayesian compressive sensing and blind deconvolution. In *Proceedings of the 20th European Signal Processing Conference (EUSIPCO)*, pages 1414–1418.

[Šroubek and Flusser, 2003] Šroubek, F. and Flusser, J. (2003). Multichannel blind iterative image restoration. *IEEE Transactions on Image Processing*, 12(9):1094–1106.

[Starck et al., 2005] Starck, J., Elad, M., and Donoho, D. (2005). Image decomposition via the combination of sparse representations and a variational approach. *IEEE Transactions on Image Processing*, 14(10):1570–1582.

[Tian and Ma, 2011] Tian, J. and Ma, K. (2011). A survey on super-resolution imaging. *Signal, Image and Video Processing*, 5(3):329–342.

[Tipping and Bishop, 2006] Tipping, M. and Bishop, C. (2006). Bayesian image super resolution. US Patent 7,106,914.

[Tropp and Gilbert, 2007] Tropp, J. and Gilbert, A. (2007). Signal recovery from random measurements via orthogonal matching pursuit. *IEEE Transactions on Information Theory*, 53(12):4655–4666.

[Tsai and Huang, 1984] Tsai, R. and Huang, T. (1984). Multiframe image restoration and registration. *Advances in computer vision and Image Processing*, 1(2):317–339.

[Villena et al., 2010] Villena, S., Vega, M., Babacan, S., Molina, R., and Katsaggelos, A. (2010). Using the kullback-leibler divergence to combine image priors in super-resolution image reconstruction. In *17th IEEE International Conference on Image Processing (ICIP)*, pages 893–896. IEEE.

[Villena et al., 2013] Villena, S., Vega, M., Babacan, S., Molina, R., and Katsaggelos, A. (2013). Bayesian combination of sparse and non-sparse priors in image super resolution. *Digital Signal Processing*, 23(2):530–541.

[Villena et al., 2009] Villena, S., Vega, M., Molina, R., and Katsaggelos, A. K. (2009). Bayesian super-resolution image reconstruction using an l1 prior. In *Proceedings of 6th International Symposium on Image and Signal Processing and Analysis, 2009. ISPA 2009.*, pages 152–157. IEEE.

[Wakin et al., 2006] Wakin, M., Laska, J., Duarte, M., Baron, D., Sarvotham, S., Takhar, D., Kelly, K., and Baraniuk, R. (2006). An architecture for compressive imaging. In *IEEE International Conference on Image Processing*, pages 1273–1276. IEEE.

[Wang et al., 2010] Wang, J., Zhu, S., and Gong, Y. (2010). Resolution enhancement based on learning the sparse association of image patches. *Pattern Recognition Letters*, 31(1):1–10.

[Yang et al., 2010a] Yang, J., Wang, J., and Li, L. (2010a). A new algorithm for passive millimeter-wave image enhancement. In *2nd International Conference on Signal Processing Systems (ICSPS)*, volume 3, pages V3–507. IEEE.

[Yang et al., 2010b] Yang, J., Wright, J., Huang, T., and Ma, Y. (2010b). Image super-resolution via sparse representation. *IEEE Transactions on Image Processing*, 19(11):2861–2873.

[Yang and Zhang, 2011] Yang, J. and Zhang, Y. (2011). Alternating direction algorithms for l1-problems in compressive sensing. *SIAM Journal on Scientific Computing*, 33(1):250–278.

[Yang et al., 2011] Yang, S., Sun, F., Wang, M., Liu, Z., and Jiao, L. (2011). Novel super resolution restoration of remote sensing images based on compressive sensing and example patches-aided dictionary learning. In *International Workshop on Multi-Platform/Multi-Sensor Remote Sensing and Mapping (M2RSM)*, pages 1–6. IEEE.

[Yin et al., 2010] Yin, W., Morgan, S., Yang, J., and Zhang, Y. (2010). Practical compressive sensing with toeplitz and circulant matrices. In *Visual Communications and Image Processing 2010*, pages 77440K–77440K. International Society for Optics and Photonics.

[Yujiri et al., 2003] Yujiri, L., Shoucri, M., and Moffa, P. (2003). Passive millimeter wave imaging. *IEEE Microwave Magazine*, 4(3):39–50.

[Zhang et al., 2010] Zhang, G., Jiao, S., Xu, X., and Wang, L. (2010). Compressed sensing and reconstruction with bernoulli matrices. In *IEEE International Conference on Information and Automation (ICIA)*, pages 455–460.

[Zheng et al., 2013] Zheng, C., Yao, X., Hu, A., and Miao, J. (2013). A passive millimeter-wave imager used for concealed weapon detection. *Progress In Electromagnetics Research B*, 46:379–397.

[Zhou et al., 2015] Zhou, X., Mateos, J., Zhou, F., Molina, R., and Katsaggelos, A. K. (2015). Variational dirichlet blur kernel estimation. *IEEE Transactions on Image Processing*, 24(12):5127–5139.

[Zibetti et al., 2011] Zibetti, M., Bazán, F., and Mayer, J. (2011). Estimation of the parameters in regularized simultaneous super-resolution. *Pattern Recognition Letters*, 32(1):69–78.

[Zomet et al., 2001] Zomet, A., Rav-Acha, A., and Peleg, S. (2001). Robust super-resolution. In *Proceedings of the 2001 IEEE Computer Society Conference on Computer Vision and Pattern Recognition(CVPR)*, volume 1, pages I–645–I–650 vol.1.

In: Compressed Sensing
Editor: J. M. Sheppard

ISBN: 978-1-53613-082-9

Chapter 3

IMAGE COMPRESSED SENSED RECOVERY USING SALIENCY-BASED ADAPTIVE SENSING AND RESIDUAL RECONSTRUCTION

Ali Akbari[1,*]***, Maria Trocan***[1] ***and Bertrand Granado***[2]
[1]Institut Supérieur d'Electronique de Paris (ISEP), Paris, France [2]Laboratoire d'Informatique de Paris 6, Université Pierre-et-Marie-Curie, Paris, France

Abstract

Compressed sensing (CS) is a recently developed mathematical paradigm wherein the signals are sampled at the sub-Nyquist rates via a linear projection onto a random basis. Under certain conditions, the signals can be exactly recovered from these incomplete samples. In this chapter, several image CS recovery methods are studied. In the first part of chapter, the concept of CS in the image processing is introduced and a brief description of a well-known and yet effective CS reconstruction algorithm, called block-based CS with smoothed-projected Landweber reconstruction (BCS-SPL), is presented. In the next part, inspired by the fact that the visually salient regions of the image are more conspicuous to the human visual system, an adaptive CS method is presented, which provides a higher recovered image quality with respect to the BCS-SPL

*E-mail address: ali.akbari@isep.fr.

algorithm. Then, we extend the concept of prediction and residual reconstruction to the image CS recovery. A prediction is produced to create a measurement-domain residual of the image to be recovered; such a residual is typically more compressible than the original one, making it more adapted to the CS reconstruction. Several methods for the prediction generation are discussed in this chapter. First, considering the strong correlation among the neighboring areas within an image, we present an image CS recovery of images by providing an intra-image prediction. Further, the local sparsity and nonlocal similarities among the image patches are implicitly exploited to generate an optimal prediction, improving the performance of the image CS recovery. Finally, in the end part of chapter, objective experimental results are given for each algorithm in terms of peak-signal to noise ratio (PSNR) and the significant gains of the presented CS recovery methods are compared to the BCS-SPL algorithm.

Keywords: Compressed sensing (CS) recovery, adaptive acquisition, residual-based CS, sparse modeling

1. Introduction

Signal models are fundamental tools for efficient signal processing, including audio recordings, natural images, video clips and medical scans, to name just a few. A signal model formulates a mathematical description of the family of interesting signals in a way which faithfully captures their behavior. Designing accurate signal models, which efficiently capture the useful characteristics of the signals, has been a crucial aim in the signal processing area for so many years and a variety of mathematical forms has been proposed by now (Mairal, Bach, & Ponce, 2014). One of the simplest and most common forms is that signals have some priori probability distribution assumed on the signal space (Elad, Milanfar, & Rubinstein, 2006, 2007). While this representation is mostly useful when employing statistical signal processing methods, many signals in practical use cannot be directly associated to a probabilistic distribution.

One efficient signal modeling methods is the sparse representation which has been proven to be effective for many signals, such as natural images, and successfully used in a wide range of applications. This model represents signals in terms of linear combinations of an underlying set, called dictionary, of elementary signals, known as atoms. The driving force behind this model is the sparsity of the representation coefficients, *i.e.*, the rapid decay of the representation coefficients over the dictionary.

The sparse signal modeling has the ability to solve inverse problems such as compressed sensing (CS) recovery, denoising, and so on. The inverse problems are indeed often difficult to solve without a priori knowledge of the data source. The sparse signal modeling provides a strong tool for an a priori model of the data source. The sparse signal modeling has led to effective algorithms for many image processing applications, such as compression (Mallat, 2008; Pennec & Mallat, 2005; Villegas, Elias, Villela, Sanchez, & Salazar, 2008; Wang, Shu, Zhang, Liu, & Gan, 2008; Beferull-Lozano & Ortega, 1999; Gurumoorthy, Rajwade, Banerjee, & Rangarajan, 2010; Mazaheri, Guillemot, & Labit, 2013; Sun, Tao, Li, & Lu, 2014; Akbari, Trocan, & Granado, 2016a; Akbari & Trocan, 2015), denoising (Molina, Nunez, Cortijo, & Mateos, 2001; Blomgren & Chan, 1998; Starck, Candes, & Donoho, 2002; Portilla, Strela, Wainwright, & Simoncelli, 2003; Matalon, Elad, & Zibulevsky, 2005; Eslami & Radha, 2006; Elad & Aharon, 2006; Mairal, Elad, & Sapiro, 2008), super-resolution (Farsiu, Robinson, Elad, & Milanfar, 2004; Li, Dai, & Shen, 2010), inpainting (Elad, Figueiredo, & Ma, 2010; He & Wang, 2014; Zhang, Zhao, & Gao, 2014), image error concealment (Akbari, Trocan, & Granado, 2016b, 2017c, 2017a; Akbari et al., 2017a), and CS reconstruction (Mun & Fowler, 2009; Zhang et al., 2012), to name just a few. The main goal of this chapter is to provide new applications of the sparse signal modeling for the CS recovery from various perspectives.

2. CS-Based Image Acquisition and Reconstruction Basics

Unlike the traditional point-by-point sampling (Nyquist sampling theorem), the CS theory provides a new approach for signal acquisition wherein the signal can be exactly reconstructed using a small number of random linear measurements, under certain sparsity conditions (Candes & Wakin, 2008; Donoho, 2006). Since most signals are indeed compressible or sparse with respect to some transform domains $\mathbf{\Psi}$, the CS has attracted a lot of attentions in many applications, including medical imaging, camera design and multimedia sensor networks, due to its potential of reduction of sampling rates, power consumption and computation complexity in the image acquisition.

Mathematically, the CS effectively acquires the real-valued signal $\mathbf{x} \in \mathbb{R}^N$, via a linear projection of the signal onto some random measurement basis $\mathbf{\Phi}$ of size $M \times N$ where $M \ll N$, i.e.:

$$Cy = \mathbf{\Phi x} + \mathbf{n}, \tag{1}$$

where $\mathbf{y} \in \mathbb{R}^M$ is the measurements vector and $\mathbf{n} \in \mathbb{R}^M$ is the system noise (usually Gaussian additive white noise). The sampling operator $\mathbf{\Phi}$, called also *measurement matrix*, is a non-invertible matrix, satisfying the Restricted Isometry Property (RIP) and being incoherent with $\mathbf{\Psi}$ (Donoho, 2006). The rate of sampling, called *subrate*, achieved by (1), is defined as $S = M/N$.

How to efficiently reconstruct the image with good recovery quality and low computational complexity is always a focused topic in the CS applications field. The fundamental concept of the CS theory states that, although the number of unknowns is larger than the number of measurements, the signal can still be exactly recovered, if it is sparse enough (Candes & Wakin, 2008; Donoho, 2006). To address this problem, there have been great efforts to develop optimal CS reconstruction algorithms which differ based on their computational complexity and also the reconstruction quality (Fowler, Mun, & Tramel, 2011; Mun & Fowler, 2009; Cartis & Thompson, 2015; S. S. Chen, Donoho, & Saunders, 2001).

Without prior knowledge on the original signal, recovering $\mathbf{x}$ from $\mathbf{y}$ is an impossible task. Furthermore, the measurement matrix $\mathbf{\Phi}$ introduces further complexity because of its lossy nature that leads to the highly ill-posed and unstable direct inversion. To cope with the ill-posed nature of the CS recovery, regularization-based techniques (Oliveira, Bioucas-Dias, & Figueiredo, 2009; Daubechies, Defrise, & DeMol, 2004; Rudin & Osher, 1994) lead to the following minimization problem:

$$C\hat{\mathbf{x}} = \arg\min_{\mathbf{x}' \in \mathbb{R}^N} \|\mathbf{y} - \mathbf{\Phi}\mathrm{x}'\|_2^2 + \lambda \mathrm{S}(\mathbf{x}'), \tag{2}$$

where $\|\mathbf{y} - \mathbf{\Phi x}'\|_2^2$ is the fidelity term and $\mathrm{S}(\mathbf{x}')$ is the regularization term, used to regularize the solution space and guide the solver towards an optimal solution. λ is a regularization parameter balancing the fidelity and regularity terms. $\mathrm{S}(\cdot)$ in (2) expresses prior assumptions about the signal behavior. Finding and modeling the appropriate prior knowledge of the signals, especially natural images, plays an important role for the performance of the regularizer and directly determines the success of the image restoration process.

The early works are based on smoothness assumptions, leading for instance to the Tikhonov regularization (Franklin, 1974) and the total variation (Rudin

& Osher, 1994; Oliveira et al., 2009). These regularization methods tend to over-smooth the images due to the piecewise constant assumption. Sparsity-based signal models form another powerful regularizes and have achieved great success for solution of the inverse problems (Molina et al., 2001; Blomgren & Chan, 1998; Starck et al., 2002; Portilla et al., 2003; Matalon et al., 2005; Eslami & Radha, 2006; Farsiu et al., 2004; Li et al., 2010; Elad et al., 2010; He & Wang, 2014; Elad & Aharon, 2006; Mairal et al., 2008; Zhang et al., 2014; Dong, Shi, & Li, 2013; Zhang et al., 2012; Dong, Zhang, Shi, & Wu, 2011; Yang, Wright, Huang, & Ma, 2010). These regularization methods are based on this fact that the requisite sparsity will exist using some sparse transform $\mathbf{x} \rightarrow \boldsymbol{\Psi}(\mathbf{x})$ and it is expected that the representation coefficients $\boldsymbol{\Psi}(\mathbf{x})$ is sparse in the sense that the sorted coefficients decay rapidly. Inspired by these characteristics of the signals, the sparsity of the representation coefficients $\boldsymbol{\Psi}(\mathbf{x})$ describes the estimated likeliness of the solution as being close to the optimal one.

Based on the sparse signal modeling, the regularization term $\mathrm{S}(\cdot)$ in (2) can be defined in way to measure the sparsity of the representation coefficients $\boldsymbol{\Psi}(\mathbf{x})$ and penalizes denser representations. In practice, the ℓ_p-norms achieve these stated goals and can be considered as the sparsity measure. For instance, the ℓ_2-norm aims to decrease the length of $\boldsymbol{\Psi}(\mathbf{x})$, while penalizing the large coefficients and giving less attention to the smaller ones. On the other hand, the family of ℓ_p-norm, where $0 \leq p \leq 1$, provides a more robust tool for measuring the sparsity. Excepting ℓ_1-norm wherein the magnitudes of all coefficients $\boldsymbol{\Psi}(\mathbf{x})$ are equally penalized, other norms penalize more the non-vanishing small coefficients, while tolerating a limited number of large ones.

Given the sparse transform $\boldsymbol{\Psi}(\cdot)$ and incorporating this model into the minimization problem (2), the recovery is achieved by solving the following optimization problem:

$$C\hat{\mathbf{x}} = \underset{\mathbf{x}' \in \mathbb{R}^N}{\arg\min} \|\mathbf{y} - \boldsymbol{\Phi}\mathbf{x}'\|_2^2 + \lambda \|\boldsymbol{\Psi}(\mathbf{x}')\|_p. \tag{3}$$

Based on this modeling, the recovery algorithm searches for a sparse set of significant coefficients, $\mathbf{c} = \boldsymbol{\Psi}(\mathbf{x}')$ that are consistent with the observed measurements $\mathbf{y}$ and the magnitude of its entries decays rapidly when sorted. Then, $\mathbf{x}$ is reconstructed by $\hat{\mathbf{x}} = \boldsymbol{\Psi}^{-1}(\mathbf{c})$, where $\boldsymbol{\Psi}^{-1}(\cdot)$ is the inverse transform of $\boldsymbol{\Psi}(\cdot)$.

Designing and constructing the sparse transform $\boldsymbol{\Psi}(\cdot)$ plays an important role for the success of the minimization problem (3). One of the basic ways for construction of the transform $\Psi(\cdot)$ is based on the concept of designing a dictionary that consists of a set of vectors or atoms. In this technique, given a set of vectors $[\mathbf{d}_k \in \mathbb{R}^N]_{k=0}^{K-1}$, a signal $\mathbf{x} \in \mathbb{R}^N$ is represented as the following linear combination of K basis,

$$C\mathbf{x} \simeq \sum_{k=0}^{K-1} c_k \mathbf{d}_k, \tag{4}$$

where the set $[c_k]_{k=0}^{K-1}$ denote the representation coefficients. The signal approximation (4) can be reformulated in a matrix form as

$$C\mathbf{x} \simeq \mathbf{D}\mathbf{c}, \tag{5}$$

where $\mathbf{c} \in \mathbb{R}^K$ is the coefficients vector. The matrix $\mathbf{D} = [\mathbf{d}_1\mathbf{d}_2 \cdots \mathbf{d}_K] \in \mathbb{R}^{N\times K}$ is called dictionary and its columns constitute the dictionary atoms. If the dictionary $\mathbf{D}$ is a basis, *i.e.*, $K = N$ and atoms are linearly independent, it is often said to be complete. The representation coefficients vector $\mathbf{c} \in \mathbb{R}^K$ is obtained via $\mathbf{c} = \Psi(\mathbf{x}) = \mathbf{D}^{-1}\mathbf{x}$. When the basis is orthogonal, $\boldsymbol{\Omega} = \mathbf{D}^{-1} = \mathbf{D}^T$, *i.e.*, each representation coefficients $\mathbf{c}$ can be computed as inner products of the signal $\mathbf{x}$ and the atoms $\boldsymbol{\Omega}$.

With the right choice of dictionary, the coefficients vector $\mathbf{c}$ is expected to be sparse. Motivated by this idea, the design of an efficient orthogonal transform was an active area of research during the last decades. The well-known discrete Cosine transform (Ahmed, Natarajan, & Rao, 1974), Short Time Fourier Transform (STFT) (Allen & Rabiner, 1977), Gabor transform (Qian & Chen, 1993), wavelet transform (Mallat, 2008), and wavelet packets (Coifman, Meyer, & Wickerhauser, 1992) are the designed transforms based on this concept. Although these dictionaries have been used in many applications due to their mathematical simplicity, their limited expressiveness of the orthogonal dictionaries leads to outweigh this simplicity. Consequently, the complete dictionaries are not well equipped for representing increasingly complex natural and high dimensional signals. For instance, the representation using the wavelet basis is optimal for piecewise smooth 1-d signals with a finite number of discontinuities (point singularities) (Mallat & Hwang, 1992). However, it loses its optimality in the representation of image signals due to existence of curve singularities in these types of signals (Rubinstein, Bruckstein, & Elad, 2010). This weakness is due to the small and fixed number of atoms ($K = N$) in the dictionary, which is the consequence of the orthogonality. On the other hand, transform reversibility imposes a strict limit on the number of atoms in the dictionary ($K = N$).

In an attempt to minimize this weakness of the orthonormal dictionaries, the design of more general over-complete dictionaries which have more atoms than the dimension of the signal, i.e. $K > N$, has been investigated. These dictionaries have a more descriptive ability to represent a wide range of interesting signals, when compared with the reversible complete dictionaries. Overcomplete dictionaries such as steerable pyramids (Simoncelli, Freeman, Adelson, & Heeger, 1992), complex wavelets (Selesnick, Baraniuk, & Kingsbury, 2005), curvelets (Candés, Demanet, Donoho, & Ying, 2006), contourlets (Do & Vetterli, 2002), surfacelets (Lu & Do, 2007) and shearlets (Easley, Labate, & Lim, 2008), as well as a wide range of trained dictionaries (Olshausen & Field., 1997; Aharon, Elad, & Bruckstein, 2006; Engan, Skretting, & Husoy, 2007) are especially advantageous for multi-dimensional signals, where invertible dictionaries lose much of their effectiveness.

Introducing the sparse signal modeling (5) in (2) leads to the following optimization problem:

$$C\hat{\mathbf{x}} = \underset{\mathbf{x}' \in \mathbb{R}^N}{\arg\min} \|\mathbf{y} - \mathbf{\Phi}\mathbf{x}'\|_2^2 + \lambda \|\mathbf{\Omega}\mathbf{x}'\|_{\ell_p}, \tag{6}$$

Of specific interest is the choice $p = 1$, as it represents the most robust convex option for the minimization problem (6), which is efficiently solved by the well-known Basis Pursuit (BP) algorithm (S. S. Chen et al., 2001). Alternatively, more specialized algorithms include least absolute shrinkage and selection operator (LASSO) (Tibshirani, 1994), gradient projection for sparse reconstruction (GPSR) (Figueiredo, Nowak, & Wright, 2007), focal underdetermined system solver (FOCUSS) (Gorodnitsky & Rao, 1997), and iterative thresholding methods (Daubechies et al., 2004; Blumensath & Davies, 2009; Elad, Matalon, & Zibulevsky, 2007; Mun & Fowler, 2009).

3. Block-Based CS Reconstruction

When the signal $\mathbf{x}$ is an image or video, the CS reconstruction methods need a large memory and computational burden due to the multidimensional nature of image data. Furthermore, a large memory is needed when storing the measurement matrix $\mathbf{\Phi}$. Low complexity of the sensing and recovery schemes play an important role in designing an imaging sensor. Block-based compressed sensing (BCS) comes to solve these issues by splitting the image into K non-overlapping blocks of size $\sqrt{N_b} \times \sqrt{N_b}$ and then, the acquisition is achieved block by block independently using the same measurement matrix $\mathbf{\Phi}_b$ for each block (Gan, 2007; Mun & Fowler, 2009). Assuming $\mathbf{x}_i$ is a vector of size N_b, representing the i-th block of the image in the vector format, the corresponding measurements $\mathbf{x}_i$ are obtained via:

$$\mathbf{y}_i = \mathbf{\Phi}_b \mathbf{x}_i, \tag{7}$$

where $\mathbf{\Phi}_b$ is the measurement matrix of size $M_b \times N_b$ $(M_b < N_b)$ (Fowler et al., 2011). This procedure can be applied directly on the entire image by creating the block-diagonal matrix $\mathbf{\Phi} = \text{diag}(\mathbf{\Phi}_b, \cdots, \mathbf{\Phi}_b)$ (Mun & Fowler, 2009).

At the receiver side, each block can be reconstructed separately using the algorithm mentioned in the previous section. However, these recovery algorithms usually create unpleasant blocking artifacts. To overcome this weakness, several BCS reconstruction algorithms have been designed (Mun & Fowler, 2009; Gao, China, Xiong, Zhou, & Wang, 2012; Trinh, Dinh, Nguyen, Jeon, & Sim, 2013). Recently, the authors in (Mun & Fowler, 2009) proposed a BCS reconstruction algorithm, called BCS with smoothed-projected Landweber reconstruction (BCS-SPL), that is based on successive projections and thresholding in the transform domain. Given an initial approximation of the image $\hat{\mathbf{x}}^{[0]}$, at each iteration k, the BCS-SPL algorithm updates the $i-$th block $\hat{\mathbf{x}}_i^{[k]}$ as:

$$\hat{\mathbf{x}}_i^{[k]} = \hat{\mathbf{x}}_i^{[k]} + \mathbf{\Phi}^T \left(\mathbf{y}_i - \mathbf{\Phi} \hat{\mathbf{x}}_i^{[k]} \right). \tag{8}$$

In the next step, all the reconstructed blocks at iteration k, *i.e.*, $[\hat{\mathbf{x}}_i^{[k]}]_{i=1}^K$, are put back into the full-image plane to update the reconstructed image $\hat{\mathbf{x}}^{[k+1]}$ using smoothing and thresholding operations. The BCS-SPL algorithm uses a smoothing operator via Wiener filtering, which reduces the blocking artifacts, and a thresholding process in the transform domain $\mathbf{\Psi}(\cdot)$, which controls the local sparsity. The BCS-SPL algorithm offers the advantage of easily allowing the utilization of a wide range of sparse transforms, such as directional transforms; the most effective one is proven by experiments to be the dual-tree discrete wavelet transform (DDWT) (Mun & Fowler, 2009). Throughout this chapter, the BCS-SPL algorithm is coupled with the DDWT as the sparsity basis. Such a procedure is iterated until $||\hat{\mathbf{x}}^{[k+1]} - \hat{\mathbf{x}}^{[k]}||_2 \leq \epsilon$. The details of the BCS-SPL algorithm is summarized in Algorithm 1. In Algorithm 1, the operator $\mathrm{R}(\cdot)$ splits

Algorithm 1 BCS-SPL

Input: $\mathbf{\Phi}_b$, $[\mathbf{y}_i]_{i=1}^{K}$, N_p, $\mathbf{\Psi}(\cdot)$
Output: $\hat{\mathbf{x}}$
Initialize: $k \leftarrow 0$ and $\hat{\mathbf{x}}^{[0]} = 0$
Repeat
  $\left[\hat{\mathbf{x}}_i^{[k]}\right]_{i=1}^{K} = \mathrm{R}\left(\hat{\mathbf{x}}^{[k]}\right)$
  For $i \leftarrow 1$ to K do
    $\hat{\mathbf{x}}_i^{[k]} = \hat{\mathbf{x}}_i^{[k]} + \mathbf{\Phi}^T\left(\mathbf{y}_i - \mathbf{\Phi}\hat{\mathbf{x}}_i^{[k]}\right)$
  End
  $\hat{\mathbf{x}}^{[k]} = \mathrm{R}^{-1}\left(\left[\hat{\mathbf{x}}_i^{[k]}\right]_{i=1}^{K}\right)$
  $\hat{\mathbf{x}}^{[k]} = \mathrm{Wiener}\left(\hat{\mathbf{x}}^{[k]}\right)$
  $\left[\hat{\mathbf{x}}_i^{[k]}\right]_{i=1}^{K} = \mathrm{R}\left(\hat{\mathbf{x}}^{[k]}\right)$
  For $i \leftarrow 1$ to K do
    $\hat{\mathbf{x}}_i^{[k]} = \hat{\mathbf{x}}_i^{[k]} + \mathbf{\Phi}^T\left(\mathbf{y}_i - \mathbf{\Phi}\hat{\mathbf{x}}_i^{[k]}\right)$
  End
  $\hat{\mathbf{x}}^{[k]} = \mathrm{R}^{-1}\left(\left[\hat{\mathbf{x}}_i^{[k]}\right]_{i=1}^{K}\right)$
  $\tilde{\mathbf{x}}^{[k]} = \mathbf{\Psi}(\hat{\mathbf{x}}^{[k]})$
  $\tilde{\mathbf{x}}^{[k]} = \mathrm{Threshold}(\tilde{\mathbf{x}}^{[k]})$
  $\hat{\mathbf{x}}^{[k+1]} = \mathbf{\Psi}^{-1}(\tilde{\mathbf{x}}^{[k]})$
Until $\|\hat{\mathbf{x}}^{[k+1]} - \hat{\mathbf{x}}^{[k]}\|_2 \leq \mu$

the image into K non-overlapping blocks and the operator $\mathrm{R}^{-1}(\cdot)$ puts back the reconstructed blocks into the corresponding positions in the reconstructed image, padded with zeros elsewhere. The BCS-SPL algorithm provides a good trade-off between the computational complexity and the reconstruction quality. Please refer to (Mun & Fowler, 2009) for more details.

3.1. Adaptive Block-Based CS Recovery

The conventional BCS scheme acquires all blocks with the same number of measurements, ignoring the subjective importance of each block to the Human Visual System (HVS). On the other hand, from the view of biological vision and scientific analysis, the visual significance of each block varies with its spatial position (Borji, Sihite, & Itti, 2012). Some regions can be more sensitive to the HVS, while others have a lower level of visual interest. Therefore, it is necessary to design an adaptive BCS scheme by joining the block-based sampling and the HVS characteristics in order to achieve a smart-human recovery performance.

In order to improve the limitations of the traditional fixed-rate BCS, several adaptive BCS from various perspectives have been proposed in the literature. In (Malloy & Nowak, 2014), the authors give a theoretical analysis of the adaptive compressed sampling and confirm that the adaptive measurements significantly outperform the non-adaptive systems in practice. The existing adaptive acquisition techniques are based on extracting local features, such as standard deviation (Noor & Jacobs, 2013; Guicquero, Verdant, Dupret, & Vandergheynst, 2015), edge counting (Guicquero, Dupret, & Vandergheyns, 2013) or estimation of the reconstruction error (Malioutov, Sanghavi, & Willsky, 2010) in the measurement domain. Inspired by relationship between compressibility and redundancy of the natural images, an adaptive scheme is proposed in (J. Chen, Zhang, & Meng, 2014) by estimation of the compressibility based on the local redundancy, measured by the statistics of the pre-sensed measurements. These adaptive samples can benefit from on-the-fly acquisition, but are often not accurate. Some few works perform the adaptive BCS by combining the sampling and reconstruction together. In (Warnell, Bhattacharya, Chellappa, & Basar, 2015), this joint sampling is used to improve the reconstruction quality for pedestrian tracking in the video surveillance applications.

3.2. Saliency-Based BCS Recovery

Visual saliency is a cognitive mechanism of the HVS in order to accurately identify the significant visual information (salient or foreground regions) and filter out other redundant visual information (non-salient or background regions) when exploring the natural scenes. Since the HVS performs numerous functions when viewing, it is not possible to combine this complex function into the direct sensing process. Beside of the hardware problems, the process of integrating the salient region detection into the cameras is not straightforward due to

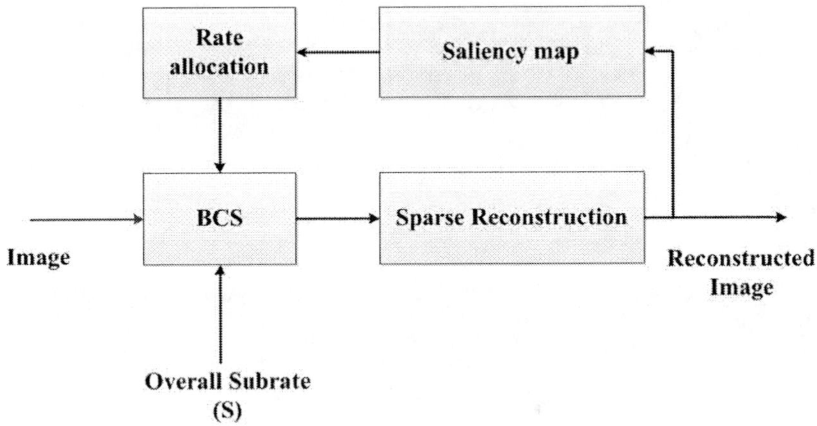

Figure 1. Block diagram of the SBCS-SPL framework.

its time- and energy- consuming process. However, by considering a feedback channel between the sensor and the receiver, the BCS can gain from the visual saliency to improve the image reconstruction quality (Akbari, Mandache, Trocan, & Granado, 2016). The graph-based visual saliency (GBVS) model (Harel, Koch, & Perona, 2006) is a well-known algorithm to extract the salient regions within the image. In the following, the GBVS and BCS-SPL algorithms are employed to build up an image CS recovery algorithm in an adaptive way to the human visual perception, called saliency-based BCS with smoothed-projected reconstruction (SBCS-SPL). Following this adaptive procedure, the high sampling rates are assigned to the salient regions while the low sampling rates are allocated to the non-salient regions, lead to a better recovery quality.

The block diagram of the SBCS-SPL scheme is shown in the Figure 1. Firstly, the basic BCS acquisition is done at the sensor side, *i.e.*, the image is partitioned into $\sqrt{N_b} \times \sqrt{N_b}$ non-overlapping blocks sampled at the same subrate S. At the receiver side, an initial reconstruction of the image is obtained using the BCS-SPL algorithm. In the next step, a saliency map for the whole image is computed using the initial reconstructed image via the GBVS model (Harel et al., 2006). Figure 2 shows some initial reconstructed images and their corresponding saliency maps at the target subrate $S = 0.3$. Brighter regions in the saliency maps represent the salient area within the test images, while the darker sections show the less-saliency regions.

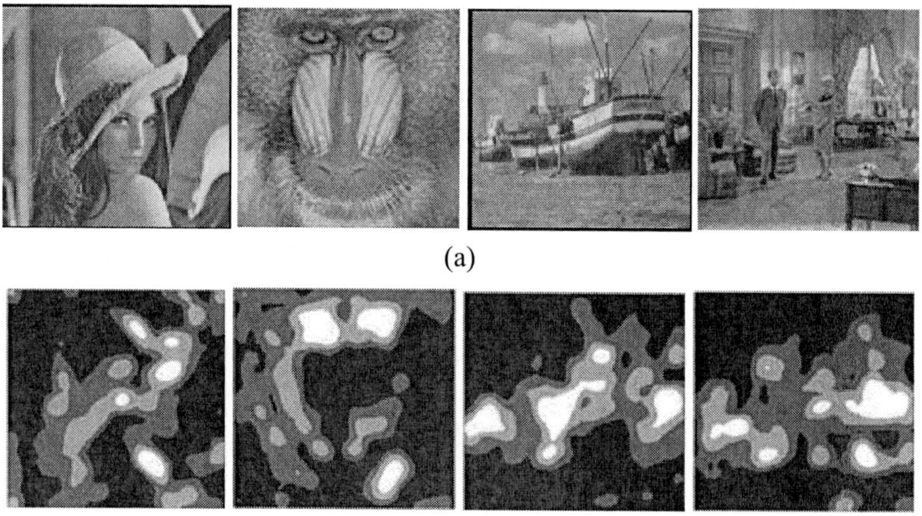

(a)

(b)

Figure 2. (a) Initial reconstructed images. (b) Corresponding saliency maps at the $S = 0.3$. The saliency maps were quantized for better display.

Based on the obtained saliency map, a binary label is assigned to each block; label 1 for the blocks belonging to the salient regions and label 0 for the blocks within the other area of the image. To this purpose, the saliency map of the input image is normalized into the range of $[0, 1]$. The pixel values of the normalized saliency map can be interpreted as the probability of that pixel belonging to the salient regions. A block is considered to be salient (labelled with 1), if the averaged probability of its pixels, called the salience value of the block, is greater than a predetermined threshold T in the range $[0, 1]$; otherwise, it is considered as a non-salient block (labelled with 0). Obviously, the percentage of blocks falling into the salient regions will be increased by decreasing the threshold. Finding the exact value of T is not a non-trivial task. In following, a simple procedure for finding an appropriate threshold T is proposed.

Following the above step, the new sampling subrates are computed based on this binary block labeling. The aim is thus to allocate different subrates to the salient and non-salient blocks, provided that the overall subrate equal (or slightly inferior) to the target subrate S. Mathematically, the subrates, assigned to the salient and non-salient blocks, *i.e.*, S_s and S_{ns} respectively, are adjusted such that:

$$N_s S_s + N_{ns} S_{ns} = (N_s + N_{ns})S, \tag{9}$$

where N_s and N_{ns} denote the number of salient and non-salient blocks, respectively. Assume the salient and non-salient subrates are a fraction of a constant value U; *i.e.*,

$$S_s = K_s U \text{ and } S_{ns} = K_{ns} U, \tag{10}$$

subject to

$$0 < K_{ns} < K_s < 1 \quad \text{and} \quad K_s + K_{ns} = 1, \tag{11}$$

where K_s and K_{ns} are two predetermined parameters being related to the proportion of the salient and non-salient blocks in the corresponding number of measurements. From (10) and (11), one can easily solve (9) for U by:

$$U = \frac{(N_s + N_{ns})S}{K_s N_s + K_{ns} N_{ns}}. \tag{12}$$

It should be noted that taking a small value for the parameter K_{ns} can lead to serious distortion in the non-salient regions, while a selection of large value for K_{ns} cannot reflect the superiority of this scheme because of the inconspicuous contribution in the adaptive sampling.

At high target subrate S, finding the subrates S_s and S_{ns} using (10) may lead to $S_s > 1$. In this case, we impose $S_s = 1$ and then S_{ns} can be easily computed from (9) via the following equation:

$$S_{ns} = \frac{(N_s + N_{ns})S - N_s}{N_{ns}}. \tag{13}$$

In contrast, the value of subrate S_{ns} obtained with (10) is very small at a low target subrate S and can result into inaccurate reconstruction of the non-salient blocks. In order to guarantee that the assigned subrate to the non-salient blocks is not too small, a minimum subrate S_m is considered for the subrate S_{ns}. We set up $S_m = K \times S$, where $0 < K < 1$ is a predefined parameter to adjust the minimum subrate for the non-salient blocks. After obtaining S_s and S_{ns} using (10), if $S_{ns} \leq S_m$, we set $S_{ns} = S_m$ and the subrate of the salient blocks is re-calculated using (9) via:

$$S_s = \frac{(N_s + (1 - K)N_{ns})S}{N_s}. \tag{14}$$

Finally, the block labels and the sampling subrates corresponding to the salient and non-salient blocks, *i.e.*, N_s and N_{ns}, are sent to the sensor in order to perform a new adaptive acquisition. As an overall result, the measurements for i-th block is achieved by $\mathbf{y}_i = \mathbf{\Phi}_b \mathbf{x}_i$, where a smaller subrate than S, *i.e.*, S_{ns}, is assigned to the non-salient regions, while the visually significant blocks are sampled at a higher subrate than S, *i.e.*, S_s, so that the overall subrate of image is unchanged. After getting the new adaptive measurements for each block, the SBCS-SPL scheme follows the same procedure with the BCS-SPL algorithm to recover the original image.

It is worth to say that each block has a different saliency value. Therefore, the quality of the reconstructed image is improved by assigning different sampling rates based on their saliency value. On the other hand, this procedure increases the overhead information to send back to the sensor which needs a high-bandwidth feedback connection. However, the SBCS-SPL scheme just sends a binary sequence and two obtained values of sampling subrates, *i.e.*, S_s and S_{ns}, to the sensor. Therefore, a low-bandwidth connection is enough for transmitting the side information to the sensor.

The number of salient and non-salient blocks, N_s and N_{ns}, depends on the predetermined value of the threshold T. Finding an appropriate threshold T is a non-trivial task. Choosing a large threshold value T leads to a small number of the salient blocks which does not have a big effect on the performance. On the other hand, with considering a small value of the threshold T, a large part of the image is considered as the salient region. To select an appropriate value for the threshold value, several experiments have been conducted on several test images by changing T from 0 to 1. At each threshold, the test images are sampled using the SBCS-SPL scheme and then the peak signal to noise ratio (PSNR) value of the reconstructed image is computed. The resulting PSNR values versus T is shown in Figure 3 for the *Lena* image at the target subrate $S = 0.3$. This curve provides a reliable comparison of how proportion of the salient regions, resulted by changing the threshold T, has an effect on the reconstructed image. As can be seen in Figure 3, the PSNR does not change proportionally with the percentage of the salient regions. As T increases, the PSNR value increases, but a larger value of T does not yield higher performance. It is easy to understand that the PSNR value first rises and then declines, when the the threshold T becomes larger. A good recovery precision can be achieved by choosing the threshold T

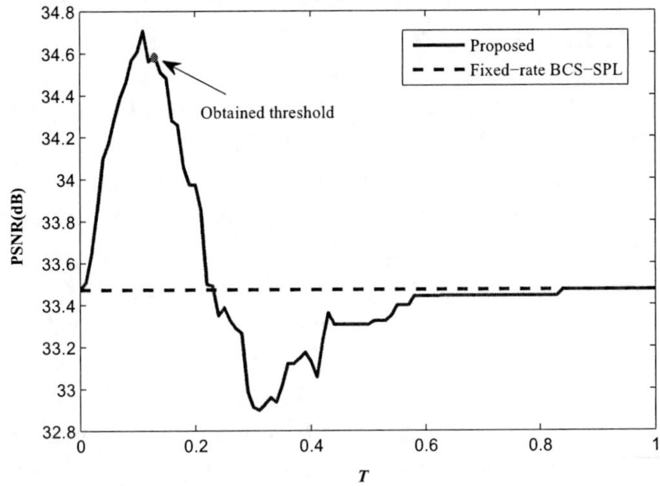

Figure 3. Effect of threshold value T on quality of the reconstructed image *Lena* in terms of PSNR at the overall subrate $S = 0.3$.

by:

$$T = \frac{1}{2(R \times C)} \sum_{x=0}^{R-1} \sum_{y=0}^{C-1} H(x, y), \tag{15}$$

where R and C are the width and height of the normalized saliency map H, respectively, and $H(x, y)$ is the saliency value of the pixel at the position (x, y). As can be observed in the Figure 3, the obtained threshold for the *Lena* image is $T = 0.13$ which is near to the real maximum of the curve. At this case, the percentage of salient regions is $\%53.9$. Given the threshold T, relatively high sampling rates are assigned to the blocks in the salient regions and low sampling rates are assigned to the rest of blocks. For instance, for the *Lena* image at the target subrate $S = 0.3$, the obtained subrates for the salient and non-salient areas are $S_s = 0.46$ and $S_{ns} = 0.11$, respectively.

4. Residual-Based Image CS Reconstruction Algorithms

In the previous section, an image BCS recovery algorithm has been presented by which the recovery quality is improved via establishing a feedback channel between the receiver and sensor in an adaptive acquisition framework. This section focuses on introducing receiver-based CS recovery algorithms without any change on the part of signal acquisition.

In video compression, motion estimation and compensation are used to support the temporal sparsity. In this technique, the first frame within a group of pictures (GOP) is firstly encoded. Exploiting the high resemblance among the frames within the same GOP, instead of fully encoding the next frame, the residual between the original frame and its prediction from the previous frame, is encoded. This resulted residual is obviously more sparse. The same encoding procedure is conducted for the multi-view image and video compression, exploiting disparity estimation and compensation methods. The fundamental property of the CS theory states that a very sparse signal can be perfectly reconstructed. Therefore, higher sparsity can be provided by estimating a prediction of that image and exploiting it for the CS reconstruction (Trocan, Tramel, Fowler, & Pesquet-Popescu, 2014; Trocan, Maugey, Fowler, & Pesquet-Popescu, 2010; Trocan et al., 2014) by implicitly increasing the signal compressibility (Tramel & Fowler, 2011). In this paradigm, during the reconstruction step, a prediction of the current block is obtained by exploiting the strong correlation among the blocks within an image. Then, the CS reconstruction algorithm is driven by the measurement-domain residual that is obtained by difference between the measurements of the current block and its prediction (Fowler et al., 2011; Mandache, Akbari, & Trocan, 2015; Akbari, Trocan, & Granado, 2017b). Suppose measurements of the block $\mathbf{x}_i$ is obtained via $\mathbf{y}_i = \mathbf{\Phi}_b\mathbf{x}_i$. Let $\tilde{\mathbf{x}}_i$ is a prediction of the block $\mathbf{x}_i$, then the measurement-domain residual of the $i-$block is obtained by $\mathbf{r}_i = \mathbf{y}_i - \mathbf{\Phi}_b\tilde{\mathbf{x}}_i$. Given the measurement-domain residuals for all blocks, *i.e.*, $\mathbf{r}$, the final reconstructed frame $\hat{\mathbf{x}}$ is computed as:

$$C\hat{\mathbf{x}} = \tilde{\mathbf{x}} + \text{BCS-SPL}(\mathbf{r}, \mathbf{\Phi}_b), \tag{16}$$

where $\tilde{\mathbf{x}}$ is the prediction image. As the residual (difference between $\mathbf{x}$ and $\tilde{\mathbf{x}}$) is more compressible (sparser) than the block itself, a better approximation of the signal is obtained by the CS recovery (Trocan et al., 2010).

4.1. Residual-Based CS Reconstruction Using Intra-Image Prediction

In this section, the basic framework of intra-image prediction is exploited to propose a novel strategy for the block-based image CS reconstruction, called block-based CS reconstruction using intra-image prediction (BCS-IIP).

First, each image is partitioned into K non-overlapping blocks of size $\sqrt{N_b} \times \sqrt{N_b}$ pixels and then the measurements are separately obtained for each block using a $M_b \times N_b$ Gaussian measurement matrix $\mathbf{\Phi}_b$ via $\mathbf{y}_i = \mathbf{\Phi}_b \mathbf{x}_i$ at the subrate $S = M_b/N_b$, where $\mathbf{x}_i$ is the vector representation of the i-th block. Block diagram of the BCS-IIP algorithm is depicted in Figure 4. The algorithm consists of two phases: initial and enhancement phase. In the initial phase, the BCS-IIP algorithm builds an initial reconstruction of the image,*i.e.*, $\hat{\mathbf{x}}$. This reconstruction can be achieved using any CS reconstruction algorithm. The BCS-SPL algorithm is employed in our implementation. Moving forward to the enhancing phase, the initial reconstructed image $\hat{\mathbf{x}}$ is partitioned into L non-overlapping blocks of size $\sqrt{N_p} \times \sqrt{N_p}$. N_p is not necessarily the same as N_b. For the prediction goal, the small block sizes improve the prediction quality, whilst in the case of CS, the bigger the block the better the reconstruction. For the former, the selection of a smaller value for the block size increases the correlation among the neighboring blocks, leading to a more accurate prediction. For the latter, the larger blocks for the acquisition provides more sparse blocks, a necessary condition for success of the CS reconstruction algorithms.

Let set $\mathbf{S} = \mathbf{z}_1, \mathbf{z}_2, \cdots \mathbf{z}_8$ consists of the 8 surrounding blocks of the current block $\mathbf{z}$ in the initial reconstruction image $\hat{\mathbf{x}}$, as shown in 5. The similarity between the block $\mathbf{z}$ and its $k-$th surrounding block is measured by sum of absolute differences (SAD) as:

$$\text{SAD}^k = \sum_{m=1}^{N_p} \sum_{n=1}^{N_p} |\mathbf{z}_k(m,n) - \mathbf{z}(m,n)|. \tag{17}$$

Using these similarity measure, the decoder can decide if the block $\mathbf{z}$ and its 8 surrounding blocks are part of the same smooth area within the image. This decision is achieved as:

$$\text{SAD}^k < T, \tag{18}$$

where the threshold T is related to the block size and the maximal intensity value within the image. T can be effectively determined by $T = B_p^2.I$, which

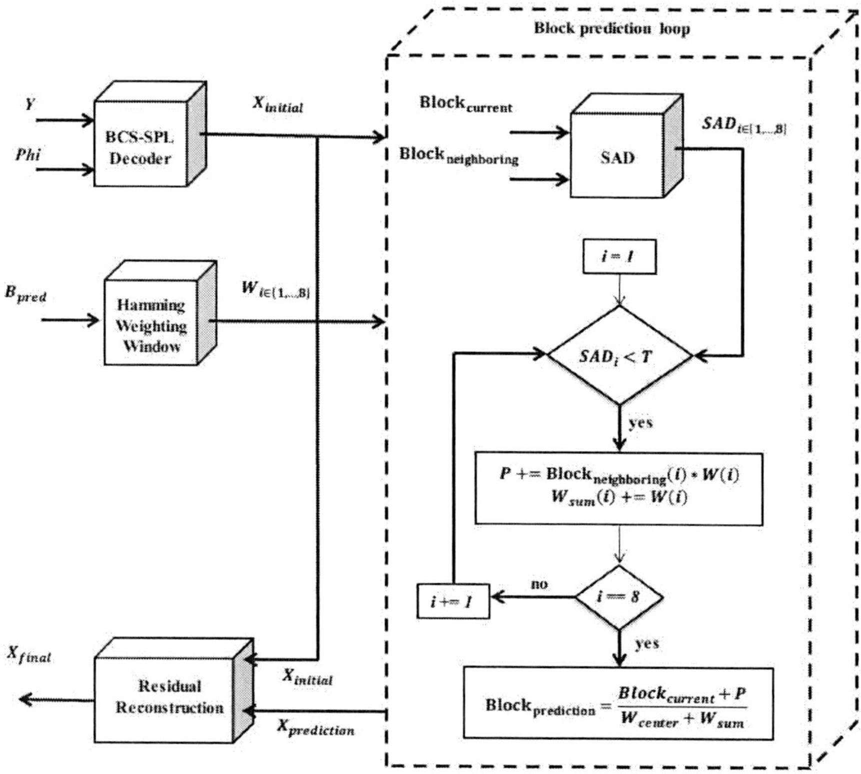

Figure 4. Block diagram of the BCS-IIP algorithm.

is equivalent to the sum of values within a block of size $\sqrt{N_p} \times \sqrt{N_p}$ pixels, whose all pixel values equals to I. Based on the SAD values and the threshold T, the neighboring blocks that satisfy the condition (18) are taken into account to generate a prediction for the current block $\mathbf{z}$.

The way of using the blocks in the set $\mathbf{S}$ influences the quality of the obtained prediction for the current block $\mathbf{z}$. A weighting scheme is used to control the influence of each sample in the resulted prediction. The weighting function allows us to emphasize the pixels which are closer to the current block and thus more important for the prediction. Consider a low-pass weighting window of size $3\sqrt{N_p} \times 3\sqrt{N_p}$ pixels, which is depicted in Figure 6. The Hamming function (Ganguly, Ganguly, & Trocan, 2014) is used for generating this weighting window. By splitting this weighting matrix into $\sqrt{N_p} \times \sqrt{N_p}$ blocks, the cor-

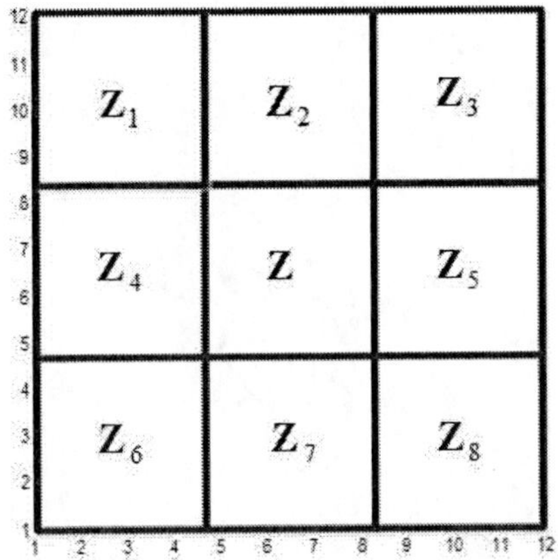

Figure 5. 8 surrounding blocks of the current block $\mathbf{z}$.

responding weighting coefficients for the 8 corresponding blocks are obtained, as shown in Figure 7. The central block in the weighting window, denoted by $\mathbf{w}_c$, is replaced with a block whose all values equal to 1, meaning that the current block $\mathbf{z}$ contributes entirely to its prediction. Finally, the prediction of the current block $\mathbf{z}$ is obtained by the following equation:

$$\mathbf{z}_p = \frac{\mathbf{z} + \sum_{k \in \mathbf{S}_s} \mathbf{z}_k * \mathbf{w}_k}{\mathbf{w}_c + \sum_{k \in \mathbf{S}_s} \mathbf{w}_k}, \tag{19}$$

where $*$ shows the element-wise matrix multiplication and the set $\mathbf{S}_s$ denotes the set of neighboring blocks which respect to (18). $\mathbf{w}_k$ is the corresponding submatrix for the block $\mathbf{z}_k$ belonging to the set $\mathbf{S}$. This prediction process is done for all the blocks within the initial image $\hat{\mathbf{x}}$ to generate a prediction $\tilde{\mathbf{x}}$ for the image $\mathbf{x}$.

With respect to the same BCS at the encoder side, the prediction image $\tilde{\mathbf{x}}$ is partitioned into $\sqrt{N_b} \times \sqrt{N_b}$ blocks and then the measurements are separately obtained for each block by $\tilde{\mathbf{y}}_i = \mathbf{\Phi}_b \tilde{\mathbf{x}}_i$. Note that the same measurement matrix should be used for sensing the blocks in the original and prediction images. Given the measurements $\tilde{\mathbf{y}}_i$, a residual in the measurement domain is obtained for each block via $\mathbf{r}_i = \mathbf{y}_i - \tilde{\mathbf{y}}_i$. Given the measurement matrix $\mathbf{\Phi}_b$ and the measurement-domain residuals $\mathbf{r} = [\mathbf{r}_i]_{i=1}^{L}$ for all blocks within the image, the spatial-domain residual image is obtained using the BCS-SPL algo-

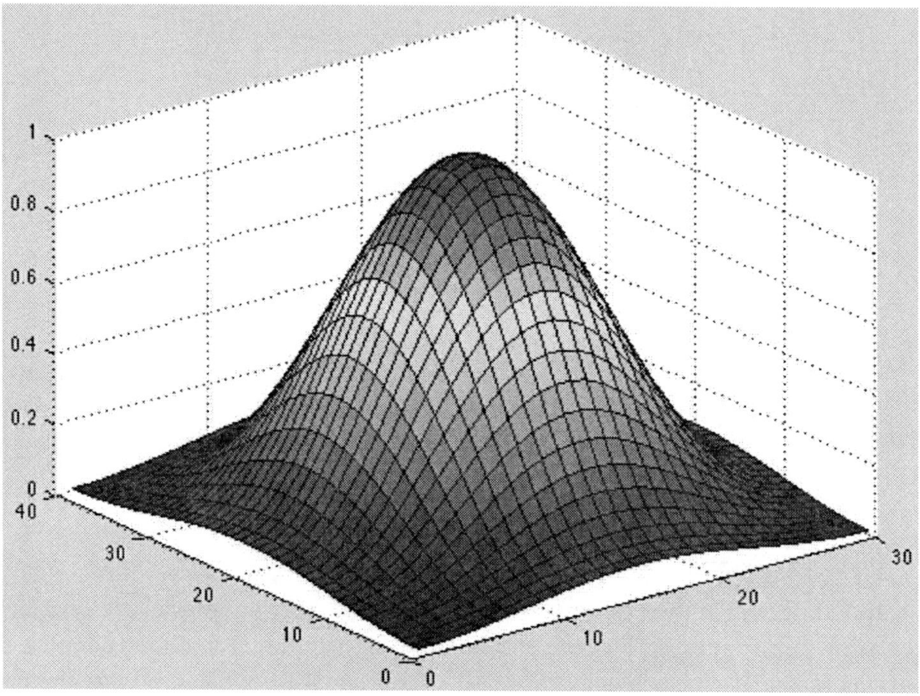

Figure 6. Hamming weighting window.

rithm. Finally, the image is obtained by combining the prediction image $\tilde{\mathbf{x}}$ and the reconstructed residual image:

$$\mathbf{x} = \tilde{\mathbf{x}} + \text{BCS-SPL}(\mathbf{r}, \mathbf{\Phi}_b), \tag{20}$$

Inspired by the CS theory, the CS reconstruction algorithms are able to recover this residual image more accurate than recovering the original image. Therefore, building a CS reconstruction based on the residual measurements and coupling it with the image prediction results in the enhancement of recovery quality.

4.2. Residual-Based Compressed Sensing Recovery Using Sparse Representations over a Trained Dictionary

In the previous section, a residual-based CS reconstruction algorithm has been described wherein the final reconstructed image is obtained using a

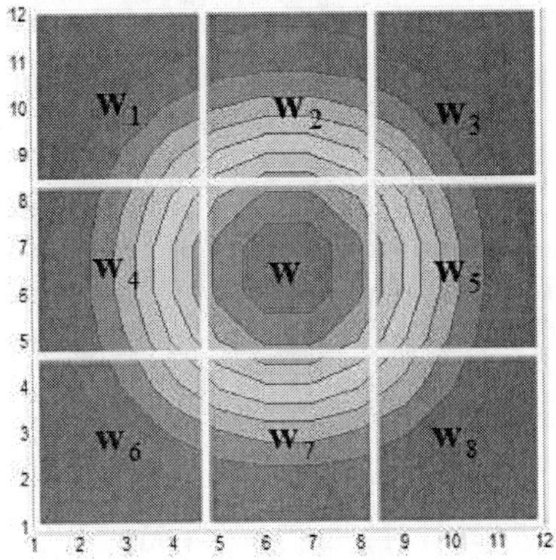

Figure 7. Corresponding weighting coefficients for 8 corresponding blocks.

measurement-domain residual. Generally, the CS recovery on the residual image, obtained by difference of the original image and its prediction, increases the CS reconstruction performance (Mandache et al., 2015; C. Chen, Tramel, & Fowler, 2011). In the BCS-IIP algorithm, the weighted average of the spatially surrounding blocks is computed to obtain a prediction for each block. However, the limitation of this method is that its performance relies heavily on the obtained prediction that is based on the assumption of similarity among the adjacent blocks within an image. To address this issue, the non-local self-similarities of image patches within an image can be exploited in order to produce a prediction that is as close as possible to each block in order to generate a highly compressible residual (Akbari et al., 2017b).

The well-known sparse representation (SR) models the image patches as a linear combination of a very few number of atoms chosen from an over-complete dictionary. Group-based sparse representation (GSR) (Zhang et al., 2014) is a novel modeling technique of the natural images. In this technique, instead of modeling the single image patches, a group of non-local image patches with similar structures is considered as the basic unit of the sparse representation. Based on this concept, the GSR is used to obtain an optimal prediction for each block. This prediction drives a residual-based CS reconstruction algorithm, called block-based CS reconstruction using group sparse representation (BCS-GSR).

Assume the image $\mathbf{x}$ of size $\sqrt{N} \times \sqrt{N}$ pixels is partitioned into K non-overlapping blocks of size $\sqrt{N_b} \times \sqrt{N_b}$ pixels. Then, each block is sampled as $\mathbf{y}_i = \mathbf{\Phi}_b \mathbf{x}_i$, where $\mathbf{x}_i \in \mathbb{R}^{N_b}$ is the vector representation of i-th block of the image, $\mathbf{y}_i \in \mathbb{R}^{M_b}$ is the measurement vector, and $\mathbf{\Phi}_b$ is the measurement matrix of size $M_b \times N_b$. Assume all block are sampled at the same subrate; therefore, the overall subrate is $S = M_b/N_b$.

Given the measurement vectors $[\mathbf{y}_i]_{i=1}^K$, the object is to recover the image patches $[\mathbf{x}_i]_{i=1}^K$ in order to reconstruct the whole image $\mathbf{x}$. In the first step, the aim is to find an optimal prediction of the image $\mathbf{x}$ by solving the following minimization problem:

$$C\tilde{\mathbf{x}} = \arg\min_{\mathbf{z}} \|\mathbf{x} - \mathbf{z}\|_2 + \lambda \mathbf{\Theta}(\mathbf{z}), \tag{21}$$

where $\|\mathbf{x} - \mathbf{z}\|_2$ is the ℓ_2 data-fidelity term, $\mathbf{\Theta}(\mathbf{z})$ is the regularization term that regularizes the solution space of the fidelity term, and λ is the regularization parameter. As discussed in the section 2, one of the most significant regularization terms is sparsity, which means that the natural images are locally sparse. In other words, each local image patch can be accurately represented as a few elements chosen from a dictionary (Rubinstein et al., 2010; Yang et al., 2010; Elad & Aharon, 2006). Further, there are many repetitive high level patterns and regular structures globally positioned in the images (Zhang et al., 2014). This significant property between similar patches within an image, called non-local self-similarity, in combination with the local sparsity existing in the natural images, can be used as a more suitable and efficient regularization term to regularize the solution space of the minimization problem in (21) and develop a much more accurate prediction (Zhang et al., 2014; Sun, Liu, Tang, & Tao, 2014; Dong et al., 2011).

Assume the image $\mathbf{x}$ is firstly partitioned into L overlapping patches of size $\sqrt{N_p} \times \sqrt{N_p}$ pixels, represented as column vectors $[\mathbf{x}_i^p]_{i=1}^L$, where $\mathbf{x}_i^p \in \mathbb{R}^{N_p}$. Assume matrix $\mathbf{H}_i \in \mathbb{R}^{N_p \times C}$ consists of C similar patches to the $\mathbf{x}_i$ within a large enough window of size $B \times B$ pixels, as shown in Figure 8. The Euclidean distance is considered as the similarity measure. Let to define the operator $\mathrm{R}_i(\cdot)$ that extracts the matrix $\mathbf{H}_i$ from the image $\mathbf{x}$, i.e. $\mathbf{H}_i = \mathrm{R}_i(\mathbf{x})$. Given all the

Figure 8. One image patch and its associated window for finding the similar patches.

matrices $[\mathbf{H}_i]_{i=1}^{L}$, the image $\mathbf{x}$ is reconstructed by:

$$C\mathbf{x} = \sum_{i=1}^{L} \mathrm{R}_i^{\mathrm{T}}(\mathbf{H}_i) \cdot / \sum_{i=1}^{L} \mathrm{R}_i^{\mathrm{T}}(\mathbf{1}_{N_b \times C}), \tag{22}$$

where the operator $\mathrm{R}_i^{\mathrm{T}}(\cdot)$ puts back the patches in the matrix $\mathbf{H}_i$ into the corresponding positions in the reconstructed image, padded with zeros elsewhere. The operator $\cdot/$ denotes the pixel-wise division and $\mathbf{1}_{N_s \times C}$ is an unit matrix of the same size of the matrix $\mathbf{H}_i$. It can be easily seen that the reconstructed image via Equation (22) is obtained by averaging all the overlapped patches. As will be seen later, this overlapping partitioning technique ensures that the obtained prediction does not suffer from the blocking artifacts and provides a high quality prediction.

The prediction in the BCS-GSR algorithm is obtained by assuming that each group of similar patches, i.e. $\mathbf{H}_i$, can be represented by linear combinations of a few of atoms chosen from an over-complete dictionary $\mathbf{D}$. Suppose

$\mathbf{D} = [\mathbf{D}_1\,\mathbf{D}_2 \cdots \mathbf{D}_M]$ whose atoms $[\mathbf{D}_j]_{j=1}^{M}$ are matrices of the same size as the matrix $\mathbf{H}_i$. With this structure, the dictionary $\mathbf{D}$ is a matrix of size $N_s \times CM$. Given the dictionary $\mathbf{D}$ with M atoms, the object is to find the sparse representation of each group $\mathbf{H}_i$ with respect to dictionary $\mathbf{D}$. In other words, we want to find a sparse vector $\boldsymbol{\alpha}_i \in \mathbb{R}^M$ such that $\mathbf{H}_i \simeq \sum_{j=1}^{M} \alpha_{i,j}\mathbf{d}_j$, where $\alpha_{i,j}$ is the j-th element of the sparse vector $\boldsymbol{\alpha}_i$. Mathematically, the sparse coding of $\mathbf{H}_i$ with respect to $\mathbf{D}$ is obtained by solving the following minimization problem:

$$C \underset{\boldsymbol{\alpha}_i \in \mathbb{R}^M}{\arg\min} ||\mathbf{H}_i - \sum_{j=1}^{M} \alpha_{i,j}\mathbf{d}_j||_2 + \lambda ||\boldsymbol{\alpha}_i||_0, \tag{23}$$

where λ is a regularization parameter. In this framework, the ℓ_0-norm of $\boldsymbol{\alpha}_i$ enforces the sparsity and the self-similarity of patches is considered in the definition of $\mathbf{H}_i$. Given the sparse representation vectors $[\boldsymbol{\alpha}_i]_{i=1}^{L}$ and by considering Equation (22), image x can be estimated as:

$$C\mathbf{x} \simeq \mathbf{D} \odot \boldsymbol{\alpha} = \sum_{i=1}^{L} \mathrm{R}_i^{\mathrm{T}} \left(\sum_{j=1}^{M} \alpha_{i,j}\mathbf{D}_j \right) \cdot / \sum_{i=1}^{L} \mathrm{R}_i^{\mathrm{T}}(\mathbf{1}_{N_p \times C}). \tag{24}$$

where $\boldsymbol{\alpha}$ denotes the concatenation of all sparse representation vectors $[\boldsymbol{\alpha}_i]_{i=1}^{L}$. The notation $\mathbf{D} \odot \boldsymbol{\alpha}$ is just for simplicity of the following equations.

By considering Equations Equation (21), (23), and (24), the reconstructed image via $\mathbf{x} \simeq \mathbf{D} \odot \boldsymbol{\alpha}$ can be considered as an optimal prediction for the image, where $\boldsymbol{\alpha}$ is obtained by solving the following minimization problem:

$$C \underset{\boldsymbol{\alpha}}{\arg\min} ||\mathbf{x} - \mathbf{D} \odot \boldsymbol{\alpha}||_2 + \lambda ||\boldsymbol{\alpha}||_0. \tag{25}$$

However, since the original image $\mathbf{x}$ is unknown, the CS recovery cannot be directly achieved. However, it can be successfully solved by applying an iterative procedure. First, an initial image, denoted by $\hat{\mathbf{x}}^{[0]}$, is reconstructed using the BCS-SPL algorithm. At each iteration k, the reconstructed image is updated through two steps as follows:

Step1-Prediction Generation: The scheme tries to find a solution for the following minimization problem:

$$C\tilde{\boldsymbol{\alpha}}^{[k]} = \underset{\boldsymbol{\alpha}}{\arg\min} ||\tilde{\mathbf{x}}^{[k]} - \mathbf{D} \odot \boldsymbol{\alpha}^{[k]}||_2 + \lambda ||\boldsymbol{\alpha}^{[k]}||_0. \tag{26}$$

Given the dictionary $\mathbf{D}$, the minimization problem (26) is efficiently solved by the algorithm proposed in (Zhang et al., 2014). Given $\tilde{\boldsymbol{\alpha}}^{[k]}$, the generated prediction at iteration k is updated by $\tilde{\mathbf{x}}^{[k]} = \mathbf{D} \odot \tilde{\boldsymbol{\alpha}}^{[k]}$.

Step2-Residual Reconstruction: In this step, the residual-based CS reconstruction algorithm is employed to recover the image. To this purpose, the obtained prediction $\tilde{\mathrm{x}}^{[k]}$ is partitioned into K $N_b \times N_b$ non-overlapping blocks $[\tilde{\mathbf{x}}_i^{[k]}]_{i=1}^{K}$, similar to the partitioning procedure that has been done at the transmitter side. Then, each block is projected onto the measurement domain by via $\tilde{\mathbf{y}}_i^{[k]} = \boldsymbol{\Phi}_b \tilde{\mathbf{x}}_i^{[k]}$. After computing the measurement-domain residuals $\mathbf{r}^{[k]} = [\mathbf{r}_i^{[k]}]_{i=1}^{K}$, where $\mathbf{r}_i = \mathbf{y}_i^{[k]} - \tilde{\mathbf{y}}_i^{[k]}$, the spatial-domain residual image is obtained using the BCS-SPL algorithm. Finally, the image is obtained by combining the prediction image $\tilde{\mathbf{x}}$ and the reconstructed residual image as:

$$C\hat{\mathbf{x}}^{[k+1]} = \tilde{\mathbf{x}}^{[k]} + \text{BCS-SPL}(\mathbf{r}^{[k]}, \boldsymbol{\Phi}_b). \tag{27}$$

The reconstruction quality is then improved by repeating the *step1* (prediction generation) and *step2* (residual reconstruction). The details of the BCS-GSR algorithm is summarized in Algorithm 2. The decoder terminates the iterations when $\|\hat{\mathbf{x}}^{[k+1]} - \hat{\mathbf{x}}^{[k]}\|_2 \leq 0.001$. However, the decoder can terminate the reconstruction algorithm early depending on its hardware abilities.

The discussion so far has assumed that the dictionary $\mathbf{D}$ is known. Choosing the dictionary $\mathbf{D}$ is an important task for finding an optimal solution to the minimization problem in (26). The scientific community has developed two main ways for designing the dictionaries (Rubinstein et al., 2010): 1- mathematical modeling and 2- training procedure (Rubinstein et al., 2010). The former tries to find the dictionary from a set of mathematical assumptions made on the family of the signals. For instance, Fourier basis is designed for optimal representation of smooth signals, while the wavelet dictionary is more suitable for piecewise-smooth signals with point singularities. The latter provides an ability to design a dictionary adapted to a specific data signal via machine-learning techniques. The basic assumption behind this approach is that the structure of the complex natural signals can be more accurately extracted directly from the data than by using a general mathematical model (Rubinstein et al., 2010). In fact, this approach replaces prior assumptions on the signal behavior with a training process which constructs the dictionary based on the observed signal properties. Compared to the mathematical dictionaries, the trained dictionaries deliver an increased flexibility and the ability to adapt to specific signals are superior in terms of representation efficiency, at the cost of a non-structured and

Algorithm 2 BCS-GSR

Input: $\mathbf{\Phi}_b$, $[\mathbf{y}_i]_{i=1}^{K}$
Output: $\hat{\mathbf{x}}$
Initialize: $k \leftarrow 0$ and $\hat{\mathbf{x}}^{[0]} = \text{BCS-SPL}([\mathbf{y}_i]_{i=1}^{K}, \mathbf{\Phi}_b)$
Repeat
  Step1: Prediction Generation
    Solve: $\tilde{\boldsymbol{\alpha}}^{[k]} = \arg\min_{\boldsymbol{\alpha}} \|\tilde{\mathbf{x}}^{[k]} - \mathbf{D} \odot \boldsymbol{\alpha}^{[k]}\|_2 + \lambda \|\boldsymbol{\alpha}^{[k]}\|_0$
    Compute prediction: $\tilde{\mathbf{x}}^{[k]} = \mathbf{D} \odot \tilde{\boldsymbol{\alpha}}^{[k]}$
  Step2: Residual Reconstruction
    Split $\tilde{\mathbf{x}}^{[k]}$ into K non-overlapping blocks: $[\tilde{\mathbf{x}}_i^{[k]}]_{i=1}^{K}$
    For $i \leftarrow 1$ to K do
      $\mathbf{r}_i^{[k]} = \mathbf{y}_i^{[k]} - \mathbf{\Phi}_b \tilde{\mathbf{x}}_i^{[k]} (i = 1, 2, \cdots K)$
    End
    Update: $\hat{\mathbf{x}}^{[k+1]} = \tilde{\mathbf{x}}^{[k]} + \text{BCS-SPL}(\mathbf{r}^{[k]}, \mathbf{\Phi}_b)$
Until $\|\hat{\mathbf{x}}^{[k+1]} - \hat{\mathbf{x}}^{[k]}\|_2 \leq \mu$

substantially more complex dictionary. In the BCS-GSR algorithm, a dictionary is trained usinf a set of training image patches collected from a number of training signals. We adopt the dictionary learning algorithm method in (Zhang et al., 2014) wherein a self-adaptive dictionary is learned.

5. Experimental Results

In this section, the performance of the SBCS-SPL, BCS-IIP, and BCS-GSR algorithms will be evaluated via a suite of simulations carried out on a set of 8-bit grayscale standard images of size 512×512 pixels, including *Lena*, *Peppers*, *Goldhill* and *Mandrill*. The quality performance is assessed by the peak signal-to-noise ratio measure (PSNR) at different subrates S, ranging from 0.1 to 0.5. The performance of these algorithms is also compared with the BCS-SPL algorithm[1]. The same measurement matrix $\mathbf{\Phi}_b$ is employed for all competing

[1]The MATLAB code of the BCS-SPL algorithm is available in authors' website: http://my.ece.msstate.edu/faculty/fowler/BCSSPL/

algorithms. Due to the randomness of the measurement matrix $\mathbf{\Phi}_b$, the PSNR values are obtained by running the the competing methods 10 times and the average results are reported as the final values.

For all the competing algorithms, size of the non-overlapped blocks for the acquisition step at the sensor side is set to 32×32 pixels. In the SBCS-SPL algorithm, the sampling subrates of the salient and non-salient blocks, *i.e.*, S_s and S_{ns}, respectively, are determined by the Equation (10) and the parameters K_{ns}, K_s, and K are set as 0.2, 0.8, and 0.2, respectively. In the BCS-IIP algorithm, size of the non-overlapped blocks for the prediction generation step is considered to 4×4 pixels and the parameter I for determination of the threshold T in (18) is set as 8. In the BCS-GSR algorithm, size of the overlapped image patches for the prediction generation step is considered to be 8×8 pixels and the overlapping depth between adjacent patches is set to 2 pixels. Further, size of the search window for finding the similar patches is set to be 40×40 pixels and number of the similar patches in each group $\mathbf{H}_i$ is set to $C = 60$.

Table 1 gives the averaged PSNR values of the reconstructed images by different CS reconstruction techniques at various subrates, ranging from 0.1 to 0.5. As can be seen, performance of the SBCS-SPL, BCS-IIP, and BCS-GSR algorithms is better than the BCS-SPL algorithm. The results of the BCS-IIP algorithm are nearly similar to the BCS-SPL algorithm, with the PSNR improvement up to 0.25 dB. The PSNR improvement by the SBCS-SPL algorithm is between 0.1 dB to 1.3 dB. It can be concluded that if the salient parts are extracted accurately, better performance of the CS reconstruction can be realized. The results indicate that the BCS-GSR achieves the best performance at all subrates and can indeed help to significantly increase the reconstructed image quality in terms of PSNR, especially at the higher subrates. The BCS-GSR algorithm achieves up to 3.50 dB higher PSNR than the BCS-SPL algorithm.

The enlarged parts of the reconstructed image *Lena* by different CS recovery algorithms are shown in Figure9 at the target subrate $S = 0.3$. It can be generally inferred that the reconstruction error of the BCS-GSR method is lower, when it compared with the BCS-SPL algorithm. As can be seen, the fine details and sharper image edges in the salient regions of the image become much clearer compared with the other methods, especially in some complex regions which contain high frequencies (strong hairs).

The computational complexity of the above-mentioned algorithms is evaluated in terms of reconstruction time (second) in Table 2 at various subrates. The experiments are conducted on a PC with an Intel(R) Xeon(R) processor at

Table 1. Comparison of quality of reconstructed images by differenct CS recovery tecnique in terms of PSNR (dB)

		Target Subrate (S)				
		0.1	0.2	0.3	0.4	0.5
Lena	BCS-SPL	28.00	31.28	33.31	35.15	36.44
	SBCS-SPL	28.83	32.44	34.61	36.28	37.76
	BCS-IIP	28.08	31.38	33.43	35.28	36.57
	BCS-GSR	**28.68**	**32.65**	**35.17**	**37.17**	**38.91**
Peppers	BCS-SPL	28.99	31.95	33.62	34.95	36.17
	SBCS-SPL	29.24	32.44	34.09	35.38	36.51
	BCS-IIP	29.15	32.18	33.86	35.19	36.41
	BCS-GSR	**30.31**	**33.45**	**35.27**	**36.66**	**37.95**
Goldhill	BCS-SPL	26.79	28.59	29.88	31.08	32.28
	SBCS-SPL	27.00	29.03	30.37	31.63	32.91
	BCS-IIP	26.84	28.71	30.05	31.28	32.50
	BCS-GSR	**27.42**	**30.34**	**32.07**	**33.93**	**35.70**
Mandrill	BCS-SPL	20.51	21.45	22.31	23.22	24.26
	SBCS-SPL	20.66	21.70	22.64	23.66	24.84
	BCS-IIP	20.53	21.48	22.33	23.23	24.27
	BCS-GSR	**20.76**	**22.98**	**24.31**	**25.98**	**26.79**

3.20 GHz and 8-GB of RAM at 64-bits operating system of WINDOWS 7 with a non-optimized MATLAB implementation. As can be seen, the SBCS-SPL method is slower than the BCS-SPL one due to the extra process of the initial reconstruction and also extracting the salient blocks. In addition, as expected, the BCS-IIP and BCS-GSR algorithms cannot be faster than the BCS-SPL algorithm, mostly due to the prediction time. However, its advantage in image reconstruction with high accuracy is interesting.

In the SBCS-SPL, BCS-IIP, and BCS-GSR algorithms, the BCS-SPL algorithm is involved as the core of the reconstruction process. Therefore, the computational burden of the recovery process of these algorithms is substantially increased using the iterative nature of the BCS-SPL algorithm. It is worthwhile to mention that, thanks to the block-based reconstruction, it is plausible to the extent that a parallel implementation of the BCS-SPL algorithm can be trivially accomplished in hardware, which is suitable for the real-time applications. In the other words, the BCS-SPL algorithm can be easily executed in a parallel manner to provide an acceptable computational cost using an array

(a) Original Image

(b) BCS-SPL

(c) SBCS-SPL

(d) BCS-IIP

(e) BCS-GSR

Figure 9. Subjective results for the image *Lena* by different CS recovery techniques at the subrate $S = 0.3$.

of multiple instructions multiple data (MIMD)-based parallel processors. In the BCS-GSR algorithm, in addition to this parallel implementation, the decoder can early terminate the iterations, depending on its computational power, to decrease the computational complexity and yet obtain a visually satisfactory image reconstruction. Finally, it should be noted that the SBCS-SPL method can be effectively and usefully employed for the pedestrian tracking in the video surveillance applications and also sampling of the low-to-normal dynamics video sequences without increasing the computational complexity. At these applications, only the first frame within a group of pictures (GOP) is used for the saliency computation and the rate allocation. Given the high resemblance among the frames within the same GOP, the adaptive sensing uses this saliency information to guide the sampling of the next frames in order to optimize the gain of new information.

Table 2. Averaged reconstruction time (*second*) at various subrates for different CS reconstruction algorithms

		Target Subrate (S)				
		0.1	0.2	0.3	0.4	0.5
Lena	BCS-SPL	21.72	10.92	6.62	4.77	3.92
	SBCS-SPL	57.62	29.45	18.24	13.65	12.55
	BCS-IIP	28.63	16.25	10.84	6.48	5.12
	BCS-GSR	166.62	155.90	151.91	149.35	148.91

Conclusion

Several new CS reconstruction algorithms have been considered at this chapter. First, a summary of the CS methodology and a brief description of the well-known BCS-SPL algorithm have been presented in the first part of this report. Inspired by the saliency based model of the visual attention, an adaptive BCS scheme has been introduced in order to enhance the reconstruction performance of the BCS-SPL algorithm. More specifically, based on a binary saliency map, obtained with a graph-based algorithm, the presented adaptive scheme enhances the recovery quality by finding an optimal subrate trade-off between the salient and non-salient areas. In the end part of the chapter, we have moved to the residual based CS reconstruction and extended the concept of prediction to the CS recovery. The idea is based on creating an optimal prediction comparable to the original image using the strong local correlation among neighboring blocks or exploiting the non-local self-similarities existing within natural images. This prediction is used to produce a measurement-domain residual which is more compressible, thus enhancing the quality of the reconstructed image. The efficiency of the above-mentioned methods has been evaluated via a wide range of experiments on several test images in comparison with the BCS-SPL algorithm.

References

Aharon, M., Elad, M., & Bruckstein, A. (2006, Nov.). K-SVD: An algorithm for designing overcomplete dictionaries for sparse representation. *IEEE Transactions on Signal Processing*, *54*(11), 4311–4322.

Ahmed, N., Natarajan, T., & Rao, K. R. (1974, Jan.). Discrete cosine transform. *IEEE Transactions on Computers*, *C-23*(1), 90–93.

Akbari, A., Mandache, D., Trocan, M., & Granado, B. (2016, July). Adaptive saliency-based compressive sensing image reconstruction. In *Proceedings of ieee international conference on multimedia expo workshops (icmew)* (pp. 1–6). Seattle, WA.

Akbari, A., & Trocan, M. (2015, Oct.). Sparse recovery-based error concealment for multiview images. In *Proceedings of ieee international workshop on computational intelligence for multimedia understanding (iwcim)* (pp. 1–5). Prague, Czech Republic.

Akbari, A., Trocan, M., & Granado, B. (2016a, Sep.). Image compression using adaptive sparse representations over trained dictionaries. In *Proceedings of ieee international workshop on multimedia signal processing (mmsp)* (pp. 1–6). Montreal, Canada.

Akbari, A., Trocan, M., & Granado, B. (2016b, Dec.). Image error concealment using sparse representations over a trained dictionary. In *Proceedings of ieee picture coding symposium (pcs)* (pp. 1–5). Nuremberg, Germany.

Akbari, A., Trocan, M., & Granado, B. (2017a, Aug.). Joint-domain dictionary learning-based error concealment using common space mapping. *Proceedings of IEEE International Conference on Digital Signal Processing (DSP).*

Akbari, A., Trocan, M., & Granado, B. (2017b, Feb.). Residual based compressed sensing recovery using sparse representations over a trained dictionary. In *Proceedings of international itg conference on systems, communications and coding (scc)* (pp. 1–6). Hamburg, Germany.

Akbari, A., Trocan, M., & Granado, B. (2017c, June). Sparse recovery-based error concealment. *IEEE Transactions on Multimedia*, *19*(6), 1339-1350.

Allen, J. B., & Rabiner, L. R. (1977, Nov.). A unified approach to short-time fourier analysis and synthesis. *Proceedings of the IEEE*, *65*(11), 1558–1564.

Beferull-Lozano, B., & Ortega, A. (1999, Oct.). Coding techniques for oversampled steerable transforms. In *Proceedings of asilomar conference on signals, systems, and computers* (Vol. 2, pp. 1198–1202). Pacific Grove, CA.

Blomgren, P., & Chan, T. F. (1998, Mar.). Color TV: total variation methods for restoration of vector-valued images. *IEEE Transactions on Image Processing*, *7*(3), 304–309.

Blumensath, T., & Davies, M. E. (2009, Nov.). Iterative hard thresholding for compressed sensing. *Applied Computational Harmonic Analysis*, *27*(3), 265–274.

Borji, A., Sihite, D. N., & Itti, L. (2012, Oct.). Salient object detection: A benchmark. In *Proceedings of ieee conference on computer vision* (pp. 414–429). Italy.

Candés, E., Demanet, L., Donoho, D., & Ying, L. (2006). Fast discrete curvelet transforms. *SIAM Multiscale Modeling & Simulation*, *5*(3), 861–899.

Candes, E. J., & Wakin, M. B. (2008, Mar.). An introduction to compressive sampling. *IEEE Signal Processing Magazine*, *25*, 21–30.

Cartis, C., & Thompson, A. (2015, Apr.). A new and improved quantitative recovery analysis for iterative hard thresholding algorithms in compressed sensing. *IEEE Transactions on Information Theory*, *61*(4), 2019–2042.

Chen, C., Tramel, E. W., & Fowler, J. E. (2011, Nov.). Compressed-sensing recovery of images and video using multihypothesis predictions. In *Proceedings of asilomar conference on signals, systems and computers (asilomar)* (p. 1193-1198).

Chen, J., Zhang, X., & Meng, H. (2014, June). Self-adaptive sampling rate assignment and image reconstruction via combination of structured sparsity and non-local total variation priors. *Journal of Digital Signal Processing*, *29*, 54–66.

Chen, S. S., Donoho, D. L., & Saunders, M. A. (2001, Jan.). Atomic decomposition by basis pursuit. *SIAM Review*, *43*(1), 129–159.

Coifman, R. R., Meyer, Y., & Wickerhauser, V. (1992). Wavelet analysis and signal processing. In *Proceedings of proceedings of wavelets and their applications* (pp. 153–178).

Daubechies, I., Defrise, M., & DeMol, C. (2004, Mar.). An iterative thresholding algorithm for linear inverse problems with a sparsity constraint. *Journal of Communications on Pure and Applied Mathematics*, *57*(11), 1413–1457.

Do, M. N., & Vetterli, M. (2002, Nov.). Contourlets: a new directional multiresolution image representation. In *Proceedings of asilomar conference on signals, systems and computers* (Vol. 1, pp. 497–501). Pacific Grove, CA.

Dong, W., Shi, G., & Li, X. (2013, Feb.). Nonlocal image restoration with bilateral variance estimation: A low-rank approach. *IEEE Transactions on Image Processing*, *22*(2), 700–711.

Dong, W., Zhang, L., Shi, G., & Wu, X. (2011, July). Image deblurring and super-resolution by adaptive sparse domain selection and adaptive regularization. *IEEE Transactions on Image Processing*, *20*(7), 1838–1857.

Donoho, D. L. (2006, Apr.). Compressed sensing. *IEEE Transaction on Information Theory*, *52*(4), 1289–1306.

Easley, G., Labate, D., & Lim, W. (2008, July). Sparse directional image representations using the discrete shearlet transform. *Applied and Computational Harmonic Analysis*, *25*(1), 25 – 46.

Elad, M., & Aharon, M. (2006, Dec.). Image denoising via sparse and redundant representations over learned dictionaries. *IEEE Transactions on Image Processing*, *15*(12), 3736–3745.

Elad, M., Figueiredo, M. A. T., & Ma, Y. (2010, June). On the role of sparse and redundant representations in image processing. *Proceedings of the IEEE*, *98*(6), 972–982.

Elad, M., Matalon, B., & Zibulevsky, M. (2007, Nov.). Coordinate and subspace optimization methods for linear least squares with non-quadratic regularization. *Applied and Computational Harmonic Analysis*, *23*(3), 346 – 367.

Elad, M., Milanfar, P., & Rubinstein, R. (2006, Sep.). Analysis versus synthesis in signal priors. In *Proceedings of european signal processing conference (eusipco)* (pp. 1–5). Florence, Italy.

Elad, M., Milanfar, P., & Rubinstein, R. (2007, June). Analysis versus synthesis in signal priors. *Inverse Problems*, *23*(3), 947–968.

Engan, K., Skretting, K., & Husoy, J. H. (2007, Jan.). Family of iterative LS-based dictionary learning algorithms, ILS-DLA, for sparse signal representation. *Digital Signal Processing*, *17*(1), 32 – 49.

Eslami, R., & Radha, H. (2006, Nov.). Translation-invariant contourlet transform and its application to image denoising. *IEEE Transactions on Image Processing*, *15*(11), 3362-3374.

Farsiu, S., Robinson, M. D., Elad, M., & Milanfar, P. (2004, Oct.). Fast and robust multiframe super resolution. *IEEE Transactions on Image Processing*, *13*(10), 1327–1344.

Figueiredo, M. A. T., Nowak, R. D., & Wright, S. J. (2007, Dec.). Gradient projection for sparse reconstruction: Application to compressed sensing and other inverse problems. *IEEE Journal of Selected Topics in Signal Processing*, *1*(4), 586–597.

Fowler, J. E., Mun, S., & Tramel, E. W. (2011, August). Multiscale block compressed sensing with smoothed projected landweber reconstruction. In *Proceedings of ieee european signal processing conference* (pp. 564–568). Barcelona, Spain.

Franklin, J. N. (1974, Oct.). On tikhonov's method for ill-posed problems. *Mathematics of Computation*, *28*(128), 889–907.

Gan, L. (2007, July). Block compressed sensing of natural images. In *Proceedings of ieee conference on digital signal processing.* (pp. 403–406). Cardiff, UK.

Ganguly, S., Ganguly, S., & Trocan, M. (2014). An overlapped motion compensated approach for video deinterlacing. In *Proceedings of international conference on computational collective intelligence (iccci)* (pp. 644–652). Seoul, Korea.

Gao, Z., China, W., Xiong, C., Zhou, C., & Wang, H. (2012, May). Relative sparsity estimation based compressive sensing for image compression applications. In *Proceedings of international symposium on photonics and optoelectronics (sopo)* (pp. 1–4). Shanghai, China.

Gorodnitsky, I. F., & Rao, B. D. (1997, Mar.). Sparse signal reconstruction from limited data using FOCUSS: a re-weighted minimum norm algorithm. *IEEE Transactions on Signal Processing*, *45*(3), 600–616.

Guicquero, W., Dupret, A., & Vandergheyns, P. (2013, Nov.). An adaptive compressive sensing with side information. In *Proceedings of asilomar conference on signals, systems and computers* (pp. 138–142). Pacific Grove, CA.

Guicquero, W., Verdant, A., Dupret, A., & Vandergheynst, P. (2015, May). Nonuniform sampling with adaptive expectancy based on local variance. In *Proceedings of international conference on sampling theory and applications (sampta)* (pp. 254–258). Washington, DC.

Gurumoorthy, K. S., Rajwade, A., Banerjee, A., & Rangarajan, A. (2010, Feb.). A method for compact image representation using sparse matrix and tensor projections onto exemplar orthonormal bases. *IEEE Transactions on Image Processing*, *19*(2), 322–334.

Harel, J., Koch, C., & Perona, P. (2006). Graph-based visual saliency. In *Proceedings of neural information processing systems (nips)* (pp. 545–552).

He, L., & Wang, Y. (2014, Dec.). Iterative support detection-based split bregman method for wavelet frame-based image inpainting. *IEEE Transactions on Image Processing*, *23*(12), 5470–5485.

Li, Y. R., Dai, D. Q., & Shen, L. (2010, July). Multiframe super-resolution reconstruction using sparse directional regularization. *IEEE Transactions on Circuits and Systems for Video Technology*, *20*(7), 945–956.

Lu, Y. M., & Do, M. N. (2007, Apr.). Multidimensional directional filter banks and surfacelets. *IEEE Transactions on Image Processing*, *16*(4), 918–931.

Mairal, J., Bach, F., & Ponce, J. (2014). Sparse modeling for image and vision processing. *Foundations and Trends®in Computer Graphics and Vision*, *8*(2-3), 85–283.

Mairal, J., Elad, M., & Sapiro, G. (2008, Jan.). Sparse representation for color image restoration. *IEEE Transactions on Image Processing*, *17*(1), 53–69.

Malioutov, D., Sanghavi, S., & Willsky, A. (2010, Apr.). Sequential compressed sensing. *IEEE Journal of Selected Topics in Signal Processing*, *4*(2), 435–444.

Mallat, S. (2008). *A wavelet tour of signal processing* (3rd ed.). Academic Press.

Mallat, S., & Hwang, W. L. (1992, Mar.). Singularity detection and processing with wavelets. *IEEE Transactions on Information Theory*, *38*(2), 617–643.

Malloy, M. L., & Nowak, R. D. (2014, Mar.). Near-optimal adaptive compressed sensing. *IEEE Transactions on Information Theory*, *60*(7), 4001–4012.

Mandache, D., Akbari, A., & Trocan, M. (2015, Nov.). Image compressed sensing recovery using intra-block prediction. In *Proceedings of ieee international conference on telecommunications forum (telfor)* (pp. 748–751). Belgrade, Serbia.

Matalon, B., Elad, M., & Zibulevsky, M. (2005). Improved denoising of images using modelling of a redundant contourlet transform. In *Proceeding of the spie* (Vol. 5914, p. 59141Y-59141Y-12).

Mazaheri, J. A., Guillemot, C., & Labit, C. (2013, May). Learning a tree-structured dictionary for efficient image representation with adaptive sparse coding. In *Proceedings of IEEE international conference on acoustics, speech and signal processing (ICASSP)* (pp. 1320–1324). Vancouver, Canada.

Molina, R., Nunez, J., Cortijo, F. J., & Mateos, J. (2001, Mar.). Image restoration in astronomy: a bayesian perspective. *IEEE Signal Processing Mag-*

azine, *18*(2), 11–29.

Mun, S., & Fowler, J. E. (2009, Nov.). Block compressed sensing of images using directional transforms. In *Proceedings of ieee international conference on image processing (icip)* (pp. 3021–3024). Cairo, Egypt.

Noor, I., & Jacobs, E. L. (2013, July). Adaptive compressive sensing algorithm for video acquisition using single pixel camera. *Journal of Electronic Imaging*, *22*(2), 22–29.

Oliveira, J. P., Bioucas-Dias, J. M., & Figueiredo, M. A. (2009, Sep.). Adaptive total variation image deblurring: A majorizationminimization approach. *Signal Processing*, *89*(9), 1683 – 1693.

Olshausen, A., & Field., D. J. (1997, Apr.). Sparse coding with an overcomplete basis set: a strategy employed by v1? *Vision Research*, *16*(4), 3311–3325.

Pennec, E. L., & Mallat, S. (2005). Bandelet image approximation and compression. *SIAM Multiscale Modeling & Simulation*, *4*(3), 992–1039.

Portilla, J., Strela, V., Wainwright, M. J., & Simoncelli, E. P. (2003, Nov.). Image denoising using scale mixtures of gaussians in the wavelet domain. *IEEE Transactions on Image Processing*, *12*(11), 1338–1351.

Qian, S., & Chen, D. (1993, July). Discrete gabor transform. *IEEE Transactions on Signal Processing*, *41*(7), 2429–2438.

Rubinstein, R., Bruckstein, A. M., & Elad, M. (2010, June). Dictionaries for sparse representation modeling. *Proceedings of the IEEE*, *98*(6), 1045–1057.

Rudin, L. I., & Osher, S. (1994, Nov.). Total variation based image restoration with free local constraints. In *Proceedings of ieee international conference on image processing (icip)* (Vol. 1, pp. 31–35).

Selesnick, I. W., Baraniuk, R. G., & Kingsbury, N. C. (2005, Nov.). The dual-tree complex wavelet transform. *IEEE Signal Processing Magazine*, *22*(6), 123–151.

Simoncelli, E. P., Freeman, W. T., Adelson, E. H., & Heeger, D. J. (1992, Mar.). Shiftable multiscale transforms. *IEEE Transactions on Information Theory*, *38*(2), 587–607.

Starck, J. L., Candes, E. J., & Donoho, D. L. (2002, June). The curvelet transform for image denoising. *IEEE Transactions on Image Processing*, *11*(6), 670–684.

Sun, Y., Liu, Q., Tang, J., & Tao, D. (2014, Sep.). Learning discriminative dictionary for group sparse representation. *IEEE Transactions on Image*

Processing, *23*(9), 3816-3828.

Sun, Y., Tao, X., Li, Y., & Lu, J. (2014, Nov.). Dictionary learning for image coding based on multisample sparse representation. *IEEE Transactions on Circuits and Systems for Video Technology*, *24*(11), 2004–2010.

Tibshirani, R. (1994). Regression shrinkage and selection via the lasso. *Journal of the Royal Statistical Society, Series B*, *58*(1), 267–288.

Tramel, E. W., & Fowler, J. E. (2011, Mar.). Video compressed sensing with multihypothesis. In *Proceedings of data compression conference* (p. 193-202).

Trinh, C. V., Dinh, K. Q., Nguyen, V. A., Jeon, B., & Sim, D. (2013, Dec.). Compressive sensing recovery with improved hybrid filter. In *Proceedings of ieee congress on image and signal processing (cisp)* (pp. 186–191). Hangzhou, China.

Trocan, M., Maugey, T., Fowler, J., & Pesquet-Popescu, B. (2010, July). Disparity-compensated compressed-sensing reconstruction for multiview images. In *Proceedings of ieee conferance on multimedia and expo (icme)* (pp. 1225–1229). Suntec City, Singapore.

Trocan, M., Tramel, E. W., Fowler, J. E., & Pesquet-Popescu, B. (2014, Sep.). Compressed-sensing recovery of multiview image and video sequences using signal prediction. *Journal of Multimedia Tools and Applications*, *72*(1), 95–121.

Villegas, O. O. V., Elias, R. P., Villela, P. R., Sanchez, V. G. C., & Salazar, A. M. (2008, Mar.). Edging out the competition: Lossy image coding with wavelets and contourlets. *IEEE Potentials*, *27*(2), 39–44.

Wang, S., Shu, Z., Zhang, L., Liu, G., & Gan, L. (2008, May). Iterative image coding with overcomplete curvelet transform. In *Proceedings of ieee congress on image and signal processing (cisp)* (Vol. 1, pp. 666–670). Hainan, China.

Warnell, G., Bhattacharya, S., Chellappa, R., & Basar, T. (2015, Nov.). Adaptive-rate compressive sensing using side information. *IEEE Transactions on Image Processing*, *24*(11), 3846–3857.

Yang, J., Wright, J., Huang, T. S., & Ma, Y. (2010, Nov.). Image super-resolution via sparse representation. *IEEE Transactions on Image Processing*, *19*(11), 2861–2873.

Zhang, J., Zhao, D., & Gao, W. (2014, Aug.). Group-based sparse representation for image restoration. *IEEE Transactions on Image Processing*, *23*(8), 3336–3351.

Zhang, J., Zhao, D., Zhao, C., Xiong, R., Ma, S., & Gao, W. (2012, Sep.). Image compressive sensing recovery via collaborative sparsity. *IEEE Journal on Emerging and Selected Topics in Circuits and Systems*, *2*(3), 380–391.

INDEX

#

A

B

C

D

E

F

G

H

I

J

K

L

M

N

O

P

Q

R

S

T

U

V

W

Z